Published in 2020 by Welbeck

An imprint of Welbeck Non-Fiction Limited,
part of Welbeck Publishing Group.

20 Mortimer Street London W1T 3JW

A CIP catalogue record for this book is
available from the British Library

ISBN 978 1 78739 447 6

Printed in Dubai

10 9 8 7 6 5 4 3 2 1

IFL SCIENCE! PUZZLE BOOK

MORE THAN **100** PUZZLES INSPIRED
BY THE **LIGHTER SIDE OF SCIENCE**

DR GARETH MOORE

WELBECK

CONTENTS

PLANTS AND ANIMALS

SPACE

TECHNOLOGY

SOLUTIONS

INTRODUCTION

The universe is a mind-boggling 13.8 billion years old, while our planet, Earth, is a relatively youthful 4.5 billion years old. A *lot* has happened in that time. IFLScience! is the source of the very best and most interesting information about all of it, beginning right from the very start with the Big Bang all the way through to the potential ways in which the universe might end – like the Big Rip or the Big Freeze.

The *IFLScience! Puzzle Book* takes some of the most fascinating scientific concepts, discoveries and theories and provides a brief introduction to, or explanation of, each of them, before testing your powers of logic and deduction with a tricky puzzle related to the scientific content. Each chapter is themed along a different subject and ends with a quiz testing your knowledge of it; some of the answers will have been mentioned earlier in the chapter, but by no means will all of them.

This book is not designed to be some heavy textbook or scientific tome – IFLScience! is all about promoting the lighter side of science, after all. Each of the concepts are only briefly touched on, and the hope is that you will then be interested enough to read further about them on iflscience.com.

For example, if you want to learn more about supermassive black holes, you can find an article about how one can kill the entire galaxy surrounding it, or where the nearest one is to the Earth (it's only 1,000 light years away!). Or – if the vastness of space isn't your thing – then perhaps the tiny, spooky world of quantum mechanics is, in which case articles describing how scientists have managed to "reverse time" inside a quantum computer, or how they have viewed two versions of reality existing at the same time might float your boat.

Either way, enjoy both the weird science and head-scratching puzzles from the world's favourite source of online science!

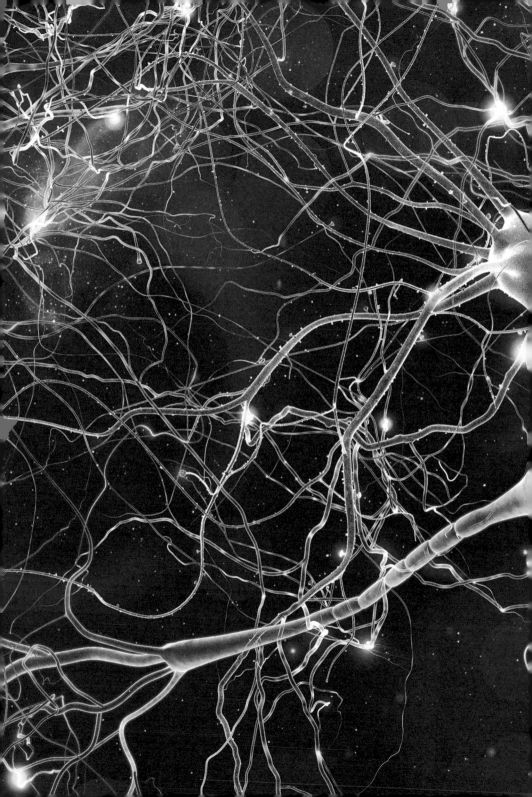

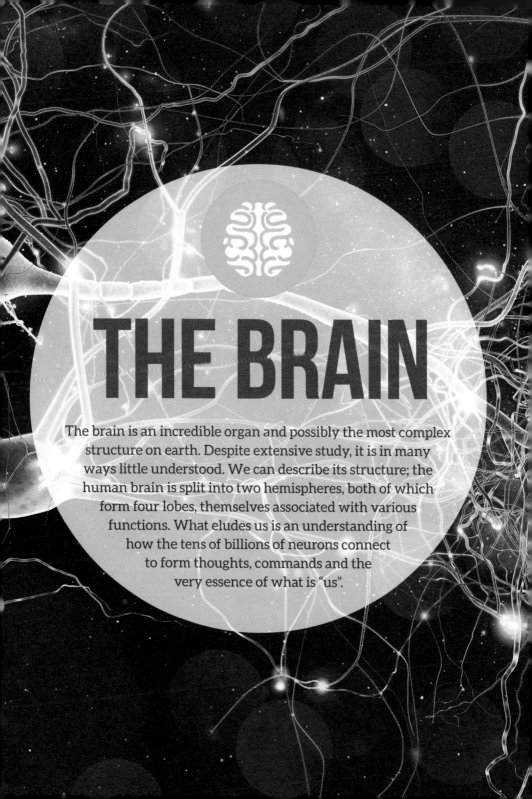

THE BRAIN

The brain is an incredible organ and possibly the most complex structure on earth. Despite extensive study, it is in many ways little understood. We can describe its structure; the human brain is split into two hemispheres, both of which form four lobes, themselves associated with various functions. What eludes us is an understanding of how the tens of billions of neurons connect to form thoughts, commands and the very essence of what is "us".

PHINEAS GAGE: THE MAN WHO HAD A ROD THROUGH HIS BRAIN

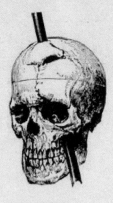

▲ An 1868 diagram of Phineas Gage's skull.
▼ Gage holding the iron rod that passed through his skull.
◢ Gage's "death mask" next to his skull displaying his head wound.

Perhaps the most famous case study in the history of neuroscience is that of Phineas Gage. In 1848, the railroad worker was injured in an accident that blew an iron rod up through his cheek and out of the top of his head. This caused severe brain damage, with the rod passing through his frontal lobes. Incredibly, he survived, but was said to have undergone an abrupt and distinctive change of character, transforming from a temperate, mild-mannered, conscientious individual to a profane, impulsive hellraiser.

Gage has since become a fixture in psychology textbooks, where he is cited as the first case study explicitly linking specific brain damage to specific changes in brain function. In Gage's case, damage to the prefrontal cortex (an area at the front of the brain) was linked to changes in the way he controlled his impulses and planned for the future. In effect, it was believed that the responsible, conscientiousness-related aspects of his personality were destroyed – functions still ascribed to this part of the brain.

In fact, most of the Gage story is now considered to be doubtful, and it is not even clear which part of his brain was injured in the accident. However, as a case study in the annals of psychology, it marked an important step on the road to linking the mind to the brain, a connection that we take for granted today but that had not been made by earlier students of the mind.

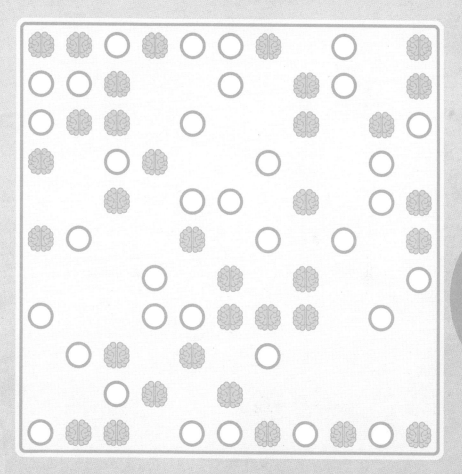

PUZZLE INSTRUCTIONS

The realisation that different areas of the brain mapped onto different personality traits and brain functions was groundbreaking. Draw horizontal or vertical lines to join the brains with the circles – representative of different brain functions – to create pairs, so that each pair contains one brain and one circle. Lines cannot cross either another line, a brain or a circle.

MIND OVER MATTER

Ancient physicians understood very well that the mind can influence the body. It is a maxim repeatedly confirmed by modern research into phenomena ranging from the effects of stress on health to the "laughter cure" research of Norman Cousins in the 1960s. Cousins suffered from a painful and apparently incurable condition: *ankylosing spondylitis*. When doctors proved unable to help him, he self-administered a diet of comedy programmes and films, and discovered that laughter soothed his pain and helped him to sleep.

This is one example of the placebo effect, which shows just how much psychology can affect physiology. The placebo effect occurs when a theoretically inactive medicine or treatment, such as a sugar pill or saline drip, produces genuine clinical outcomes like pain relief. The opposite of the placebo effect is the nocebo effect, where something that ought to be neutral and harmless has harmful effects. The nocebo effect is itself linked to various intriguing psychological phenomena, such as the voodoo effect. An individual's belief in the power of a voodoo curse produces extreme physical and mental consequences. Some believe that, and act as if, they have been turned into a zombie. Another famous example is the Baskerville effect, named after the Sherlock Holmes story, where superstitious belief or magical thinking causes death.

▼ Norman Cousins, pictured in 1974.

2	2	2			1	
				2	3	2
4			3			
2		3				1
	1		3		2	
	1			2		1
		1		2		1

PUZZLE INSTRUCTIONS

Identify which squares in the grid are contaminated by nocebo "mines". Place these mines into some empty cells in the grid. Clues in some cells show the number of mines in touching cells – including diagonally. No more than one mine may be placed per cell.

THE BEAST WITHIN

THE BRAIN

KEY

1. Frontal Lobe
2. Limbic Lobe
3. Parietal Lobe
4. Occipital Lobe
5. Corpus Callosum
6. Thalamus
7. Optic Schiasma
8. Hypothalamus
9. Pituitary Gland
10. Mammillary Body
11. Pons
12. Medulla Oblongata
13. Spinal Cord
14. Cerebellum
15. Pineal Gland

The different parts of our brain reflect the evolutionary history of humanity. The outermost layers, such as the cerebral cortex, are the parts that evolved most recently. It is no coincidence that these relate to the functions we think of as most "human", like thinking, planning and speaking. The part of the brain with the longest evolutionary history is the brainstem, where the spinal cord swells as it comes up through the neck to join the base of the brain. This part controls the simplest bodily functions, like breathing and body temperature.

In between are parts of the brain that are also found in animals such as crocodiles or wolves. In humans, these parts are collectively known as the limbic system, and they control primitive urges like lust, hate and fear. Some senses – most notably smell – stimulate the limbic system directly and can invoke strong emotional responses.

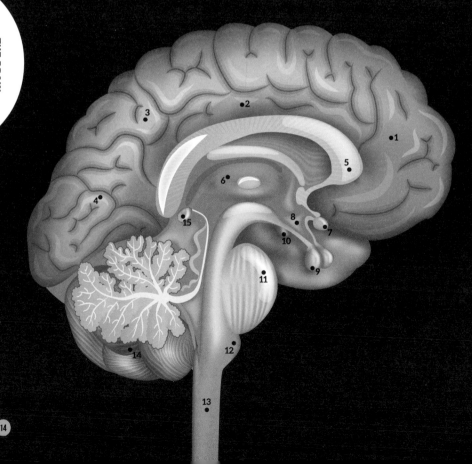

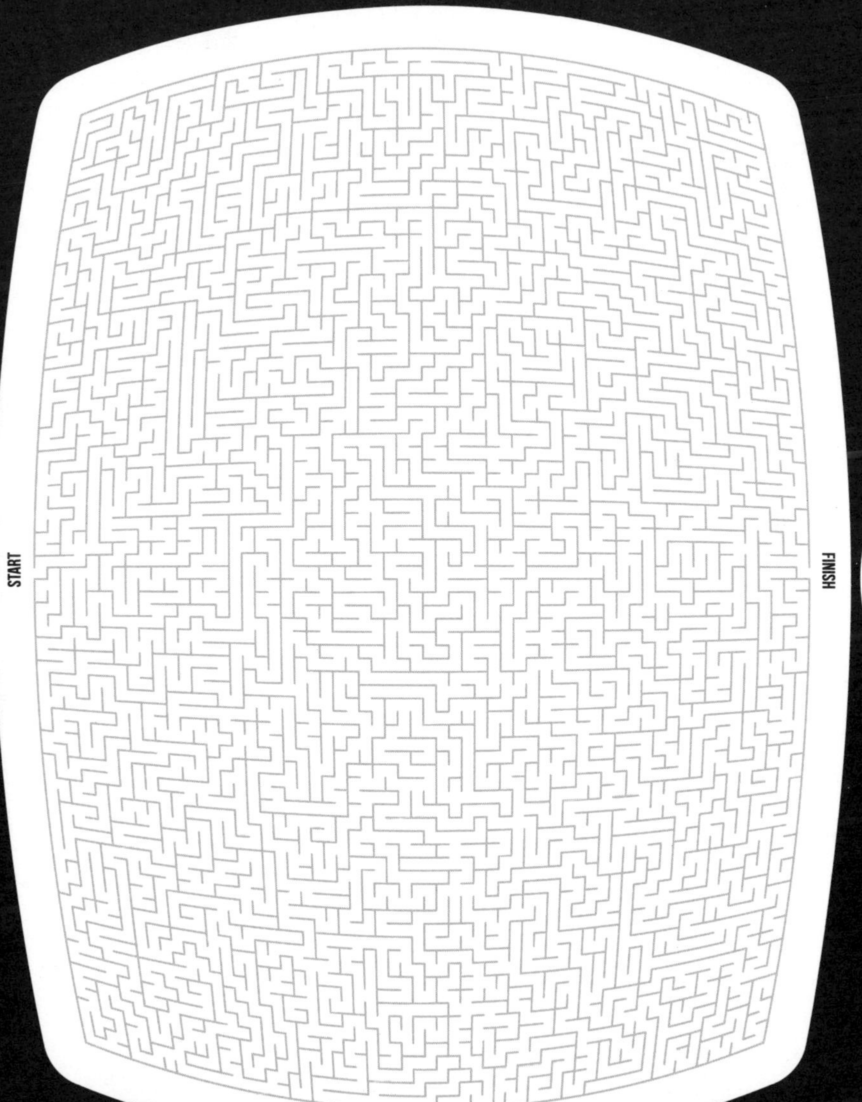

START

FINISH

PUZZLE INSTRUCTIONS

Travel like a thought through the complex maze of the human brain. Find your way from the entrance at the left to the exit at the right.

THE BRAIN OLYMPICS

Humans assume that we have the best brains in the animal kingdom, but are they the largest?

The evidence is clear. Humans certainly don't have the biggest brains; perhaps not surprisingly, these belong to the largest creatures. The biggest brain overall is that of the sperm whale – it weighs around 8 kg (18 lb), compared to c.1.4 kg (3.1 lb) for a human brain. An elephant's brain is around 5 kg (11 lb) and a bottlenose dolphin's is around 1.6 kg (3.5 lb).

We don't even triumph when it comes to the ratio of brain size to overall body size. Our ratio of approximately 1:40 is the same as that of mice, while the record holders are some types of small ant, which have a ratio of just 1:7. In other words, up to 15 per cent of their entire mass is brain. A person with a similar ratio would have a brain the size of a watermelon!

▲ An elephant never forgets... thanks to its 5 kg brain.
▼ The proud owner of the heaviest brain on earth: the sperm whale.

PUZZLE INSTRUCTIONS

Put your inferior-sized brain to the test with this ratio-based Sudoku. Place 1-9 once each into every row, column and bold-lined 3×3 box. Also, all pairs of adjacent cells where the value in one cell is equal to the value in the other cell multiplied by an integer number are marked with that integer number in a circle. For example, a circled 3 between two cells indicates that the only possible fits for the two cells are 1 & 3, 2 & 6 or 3 & 9.

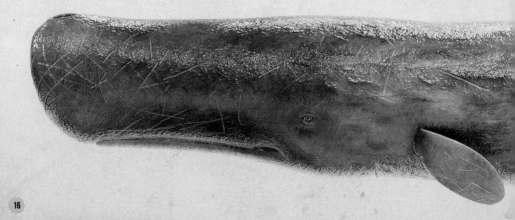

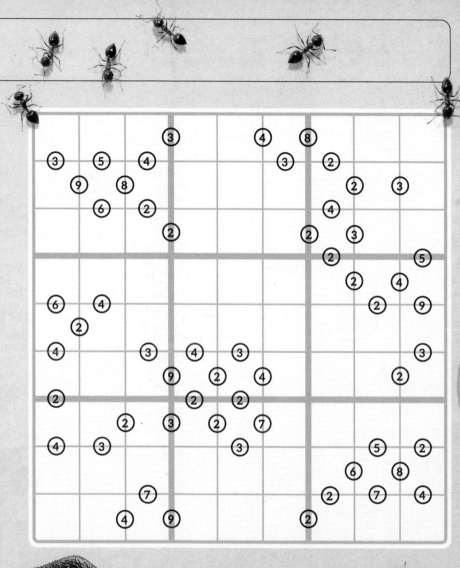

MARS VS VENUS

Men have, on average, slightly bigger brains than women, with an average difference of around 150 g. The average male neocortex (the part of the brain responsible for reasoning, memory, perception and language) has around 23 billion neurons, whereas the average female brain has around 19 billion. What difference do these extra 4 billion neurons make?

Apparently, none. Average male and female IQ scores are exactly the same. However, the detailed picture is more nuanced than this; there may be up to 100 differences between male and female brains, but they tend to be extremely minor. For example, the limbic cortex, associated with the regulation of emotions, may be slightly larger in women, while the parietal cortex, associated with spatial perception, may be slightly larger in men.

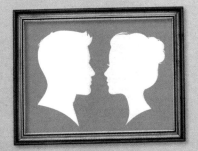

PUZZLE INSTRUCTIONS

Place an M for Mars or V for Venus in every empty cell so that there are an equal number of each letter in every row and column. Reading along a row or column, there may be no more than two of the same letter in succession. For example, MVVMMVVM would be a valid row but MVVVMMVM would not be valid due to the three Vs in succession.

V		M	V			M	
	M					V	
						V	V
	V		M		M		V
V		V				V	
M	V					V	
	M					M	
	V			M	V		M

NEURAL NETWORKS IN MAN AND MACHINE

A neural network is a system of connected neurons (which are nerve cells), or their artificial equivalents. In biology, neural networks allow complex processing to be achieved by simple units acting together. This helps to explain how amazing phenomena such as language and consciousness can emerge from a seemingly simple mass of pinkish-grey tissue.

In computer science, artificial neural networks, or "nets", are a key tool in the attempt to develop artificial intelligence. Neuron-mimicking microchips – which are effectively virtual versions of neurons – are connected together to produce a network. This network is then able to give itself feedback to strengthen or weaken its inter-connections, according to the inputs provided to, and outputs desired from, the system. What is both exciting and scary about such nets is that they can learn to perform complex tasks such as facial recognition or language processing without their human designers knowing how they work. This makes the AI network a black box that we can no longer understand.

PUZZLE INSTRUCTIONS

Join the circled numbers with horizontal or vertical lines to create a working neural network. Each number must have as many lines connected to it as specified by its value. No more than two lines may join any pair of numbers, and no lines may cross. In order to create consciousness, the finished layout must connect all numbers, so you can travel between any pair of neurons by following one or more lines.

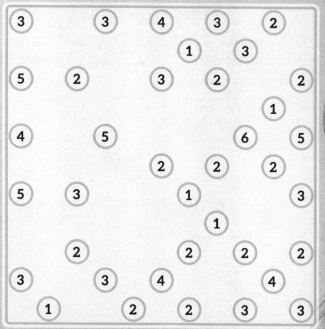

Puzzle grid (circled numbers):

Row 1: 3 · 3 · 4 · 3 · 2
Row 2: · · · · 1 · 3
Row 3: 5 · 2 · 3 · 2 · · · 2
Row 4: · · · · · · 1
Row 5: 4 · 5 · · · 6 · 5
Row 6: · · · 2 · 2 · 2
Row 7: 5 · 3 · · 1 · · · 3
Row 8: · · · · 1
Row 9: · 2 · · 2 · 2 · 2
Row 10: 3 · 3 · 4 · · · 4
Row 11: 1 · · 2 · 2 · 3 · 3

THE MAN WHO MISTOOK HIS WIFE FOR A HAT

A strange case of a man suffering from highly specific brain damage gave neurologist Oliver Sacks the title for one of his most famous books: his 1985 collection of case studies called *The Man Who Mistook His Wife for a Hat and Other Clinical Tales*. The unfortunate man had developed visual agnosia, a dysfunction in the neurological processes of perception, which left him unable to distinguish between familiar faces and everyday objects.

Damage such as strokes or head injuries can knock out specific parts of the brain, with very strange effects. For instance, people suffering from the condition known as Wernicke's aphasia, where there is damage to the part of the cortex that assigns meaning to words, speak in fluent-sounding sentences that have little or no content or meaning. This is known as "word salad".

PUZZLE INSTRUCTIONS

Find the listed "hellos" written in the word salad of a grid. The words can be in any direction, including diagonally. They may also be written backwards.

AHOJ
BONJOUR
BULA
CIAO
HALLO
HALO
HELLO
HOLA
JAMBO
KONNICHIWA
MARHABA
MERHABA
NAMASTE
NI HAO
SAIN BAINUU
SALAAM
SALVE
SANNU
SVEIKI
SZIA
XIN CHAO
ZDRAVEITE
ZDRAVO
ZDRAVSTVUYTE

```
D H S A L A A M E R H A B A H
Z A A A L U L O A L U B I Z A
E D H A N E U O A A H Z H H E
T N R T E N M N H H S I E Z S
O A K A T T U R I O C L H I B
A A E O V O I A I A L N H O I
H M B T N S B E Z O B L I T K
I A R A S N T S V H L N O X I
N B M W H A I V A A A A I O E
M O Z A E R M C U S R L H A V
O N D L O V A A H Y R D L N S
O J R A B N L M N I T E Z O O
E O A A M L M A S H W E I A N
B U V A A T O Z S A S A H O J
A R O O J O Y Z O A I C N A O
```

THE BRAIN

NO LIMITS

Forgetting is a normal – maybe even crucial – aspect of human memory, but there are intriguing hints that it may some day be possible to remember everything. Can memory really be limitless? In one of the most famous case histories in psychology, Soviet neuropsychologist Alexander Luria's 1968 classic, *The Mind of a Mnemonist*, this is indeed suggested.

Luria told the tale of a man who could remember everything, whom he called "Mr S", and is now known to have been journalist Solomon Veniaminovich Shereshevsky. Mr S had a condition called synaesthesia, where senses cross-stimulate. This condition means that a stimulus such as a sound, for example, can trigger a touch sensation, or alternatively numbers can have strong colour associations. The highly enriched sensory stimuli that Mr S experienced made incoming information easily memorable, and he proved able to recall faultlessly any material that was presented to him. Mr S, however, found his mega-memory burdensome. An unsurprising fact, as it is generally assumed that forgetting has evolved for a reason – probably to help us focus on important memories that improve our evolutionary "fitness" and limit distraction.

PUZZLE INSTRUCTIONS

Spend one minute studying the twelve objects on this page. Once time is up, turn over the page. Can you identify which objects have been replaced, and with what?

WHY DO WE DREAM?

Despite being the object of study since antiquity, and the subject of serious scientific research since the 1950s, we still don't know why we dream. There are many theories, but few seem to fit all the evidence. For instance, the popular psychodynamic theory of dreaming, with its roots in the theories of Sigmund Freud, sees dreams as the mind's way of resolving repressed desires, fears, anxieties, etc. In this way, dreaming is a sort of "mental housekeeping", or perhaps even a form of as-you-sleep self-therapy. The problem with this theory is that 99 per cent of dreams are forgotten on waking, and if dreams are supposed to be therapeutic, would it not help to remember them?

A modern spin on the Freudian approach views dreaming as a kind of virtual-reality simulator, in which the brain role-plays scenarios, presumably to improve our responses to challenging scenarios in real life. This might explain why the vast majority of dreams concern emotional content, especially negative emotions.

PUZZLE INSTRUCTIONS

Can you solve this Clouds puzzle, or is your head too up in the clouds to manage it? Shade rectangular groups of cells, such that each group is at least two cells wide and at least two cells tall. Rectangular groups of cells cannot touch, not even diagonally. Given number clues specify the number of shaded cells in each row and column.

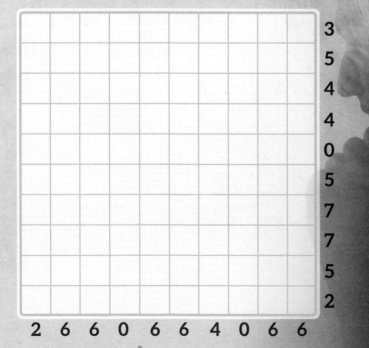

Row clues (top to bottom): 3, 5, 4, 4, 0, 5, 7, 7, 5, 2

Column clues (left to right): 2, 6, 6, 0, 6, 6, 4, 0, 6, 6

THE INVISIBLE GORILLA

The link between the outside world and your perception of it is not as straightforward as it seems. Sensory perception is only the first stage in the construction of your personal representation of the world. Much of the information that comes in through your senses is filtered out by mechanisms such as attention; you only see what you pay attention to.

A classic illustration of this was shown with an experiment in which subjects were asked to watch a video of a basketball game and count passes between players. Half of the people tested were so busy paying attention to the task that they failed to notice a man in a gorilla suit walking across the court. This "invisible gorilla" was an artefact of what is known as inattentional blindness.

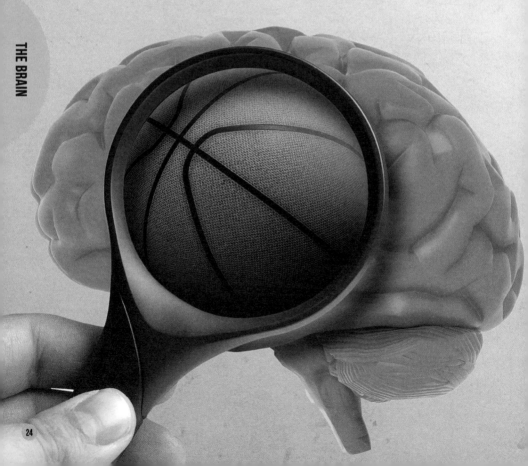

PUZZLE INSTRUCTIONS
There is more to this image than first meets the eye. List everything you see in this image.

THE BRAIN QUIZ

1) How many hemispheres does the human brain have?

2) What is the name of the outermost part of the human brain?

a) the limbic system ☐

b) the brainstem ☐

c) the cerebral cortex ☐

3) Is brain size correlated with intelligence?

4) What is the name given to the phenomenon where a person feels like they have seen something before?

5) Roughly how many neurons does a human brain contain?

a) 8.6 billion ☐

b) 86 billion ☐

c) 860 billion ☐

6) What is the branch of psychology that studies thinking, memory and language?

a) differential psychology ☐

b) cognitive psychology ☐

c) abnormal psychology ☐

7) What does REM stand for?

8) What were the three parts of Sigmund Freud's tripartite theory of personality?

9) Which of the following did Freud NOT class as a phallic symbol?

Zeppelin ☐   hanging ☐

lamp ☐   pistol ☐

fish ☐   necktie ☐

umbrella ☐

10) The Clever Hans phenomenon is where cues unconsciously exchanged between participants trigger conditioned responses. What was Clever Hans?

11) Which of these brain disorders involves movement and co-ordination?

a) Ataxia ☐

b) Alexia ☐

c) Aphasia ☐

12) Three of the most common phobias in the UK are brontophobia (fear of thunder), cardiophobia (fear of the heart and heart conditions) and aeronausiphobia (fear of air sickness). What is the most common phobia?

13) Koro is an example of a type of psychogenic illness known as a culture-bound syndrome, which involves fear of magical influence causing shrinkage or disappearance of what type of organ?

14) What delusions characterize the condition known as Fregoli syndrome?

a) Your loved ones have been replaced by impostors who look exactly like them. ☐

b) Your "real" parents are rich/famous/illustrious and will eventually reclaim you. ☐

c) Multiple different people are actually the same person, who is constantly changing their disguise or appearance. ☐

15) What is the Baskerville effect?

a) superstitious belief or magical thinking that causes people to sicken and die ☐

b) misperception of harmless phenomena as lethal ones ☐

c) delusions of being a detective ☐

16) What part of the nervous system causes blushing and goosebumps?

17) What is the name given to the phenomenon whereby a patient is cured of illness by a sugar pill believed to be a potent medicine?

18) What is the name given to the phenomenon whereby a person is made ill by what they believe to be a poison, even though it is harmless and inactive?

19) Which lobe of the brain is associated with vision?

20) By what popular name is parapraxis better known?

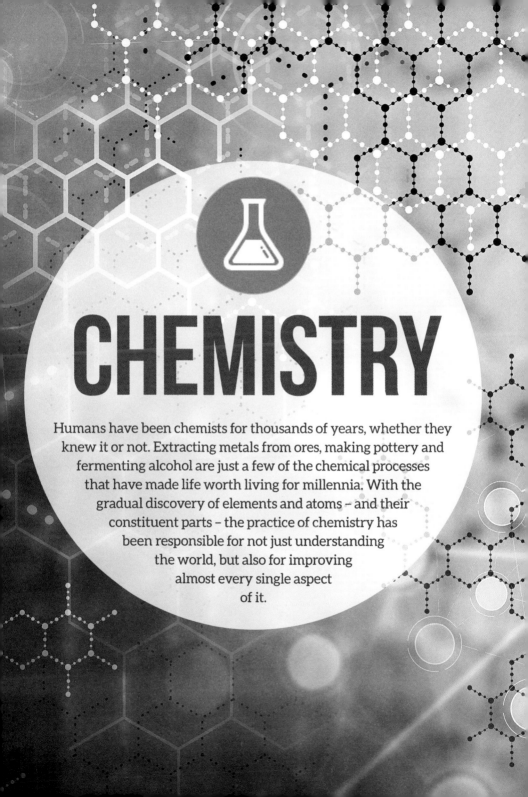

CHEMISTRY

Humans have been chemists for thousands of years, whether they knew it or not. Extracting metals from ores, making pottery and fermenting alcohol are just a few of the chemical processes that have made life worth living for millennia. With the gradual discovery of elements and atoms – and their constituent parts – the practice of chemistry has been responsible for not just understanding the world, but also for improving almost every single aspect of it.

THE QUEST FOR THE PHILOSOPHER'S STONE

The alchemists of the Middle Ages believed that they could discover the secret of transmutation of the elements. For the uninitiated, this meant transforming one element into another. They also believed in the existence of a mysterious substance – the Philosopher's Stone – which could catalyze this feat. The story went that it was able to transmute base metals, such as lead, into noble ones, such as gold.

Great scholars, from Paracelsus and John Dee to Robert Boyle and Isaac Newton, devoted tremendous effort to this project. However, by the late nineteenth century, scientific chemistry declared that their labours had all been in vain. Elements, by definition, could not transmute. This would require adding or removing protons from the nucleus, which was declared to be impossible. Even as this new doctrine was proclaimed, however, discoveries in radioactivity were serving to undermine it. It is now known that naturally occurring radioactive decay causes transmutation, and that this process can also occur inside stars and supernovae, as well as in particle accelerators (also called "atom smashers").

CHEMISTRY

	7					1
		8		3		
	7				4	
5				6	5	9
1						
		7	5		5	8
	2				3	9

PUZZLE INSTRUCTIONS
The internal division of elements perplexed scientists for centuries. See if you can divide this grid quicker than they could. Draw along the grid lines to divide the entire grid into regions, so every region contains two numbers. Each region must contain a number of grid squares that is strictly between the value of the two numbers. So, if the numbers are 5 and 12 then the region must contain between 6 and 11 squares.

DECODING SCIENTIFIC NOTATION

Alchemists had used a bewildering variety of symbols, abbreviations and deliberately obscure codes when writing about chemistry, but the new, scientific chemistry of the Enlightenment age attempted to bring to chemistry the same kind of rigour and clarity that mathematics enjoyed. The leading light of early nineteenth-century chemistry was the Swede, Jöns Jacob Berzelius (1779–1848). He insisted on "the necessity of signs" to facilitate universal transmission of chemical knowledge. It was he who inaugurated the modern system of chemical notation, in which the elements are denoted by one- or two-letter symbols derived from their Latin names. This system allowed chemical reactions to be expressed as algebra-like equations.

In the modern version of chemical notation, subscripts after the element symbol represent the number of atoms of that element in the molecule; superscripts are used to represent positive or negative charge in the case of ions. Sub- and superscripts before the element symbol represent atomic number and mass, respectively. The atomic-mass superscript, thus, also indicates the isotope of the element, so 12C, for instance, is carbon-12.

Find out how much a mole of each of these mystery elements, A, B, C and D weighs:

- The combined weight of a mole of **D** is equal to the combined weights of a mole of **A** and **B**
- Three moles of **B** weigh 1g more than one mole of **D**
- The combined weight of one mole of **A** and one mole of **C** is equal to three times the weight of a mole of **B**
- Three moles of **C** weigh 4g more than one mole of **D**
- Two moles of **A** weigh the same as three moles of **C**

A= ☐   B= ☐

C= ☐   D= ☐

A FEW DEFINITIONS

- **Atomic Mass:** the mass of an atom of a chemical element. It is approximately equivalent to the number of protons and neutrons in the atom.
- **Atomic Number:** the number of protons in the nucleus of an atom. This number gives the element its characteristics, and places it in the Periodic Table
- **Ions:** an atom or molecule with a positive or negative electric charge – usually because of the loss or gain of an electron
- **Isotope:** a form of an element that contains a different number of neutrons, e.g. carbon-12 has 12 neutrons, while carbon-14 has 14. The presence of different isotopes forms the basis of radiocarbon dating.
- **Mole:** the mass of a substance containing the same number of fundamental units as there are atoms in exactly 12g of 12C.
- **Molecule:** a group of atoms bonded together, representing the smallest fundamental unit of a chemical compound that can take part in a chemical reaction, e.g. H_2O or O_2.

THE ANGRY BEE IN THE SHRINKING BOX

Have you ever wondered why you don't fall through the floor? Or to put it another way, how things can be solid when we know that 99.9999999 per cent of an atom is empty space? The answer is that atoms have electrostatic charges, thanks to the negatively charged electrons that orbit the nucleus. These charges repel one another so that atoms cannot pass through one another.

In terms of electrostatic fields, atoms "appear" to one another to be solid, or at least enclosed in impenetrable shells. But this raises a further question: why do the negatively charged electrons remain in their orbits, distant from the nucleus, when they are attracted by the positively charged protons in the nucleus? Classical physics theories would demand this to be the case.

However, quantum mechanics has other ideas. Heisenberg's uncertainty principle describes how we cannot be certain about both the position and velocity of an electron at the same time. If an electron were to collapse into the nucleus, its position would become increasingly constrained. In turn, this would make its velocity increasingly erratic. You could imagine it like a bee trapped in a box, which buzzes more angrily the more the box shrinks, and thus refusing to "fall" into the nucleus

KEY
1. Electron
2. Proton
3. Neutron
4. Nucleus

PUZZLE INSTRUCTIONS
Locate these electrons in each region by placing one electron in some cells so that each row, column and outlined region contains exactly one electron. However, electrons cannot touch each other – not even diagonally.

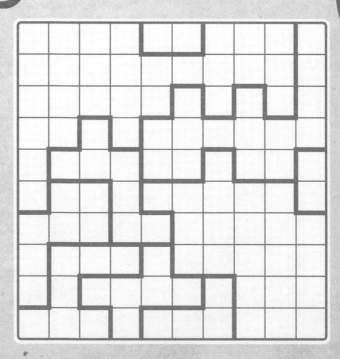

THE PERIODIC TABLE

The periodic table is a chart of all known elements, arranged according to their nuclear properties, which in turn produce their chemical properties. The Russian chemist Dimitry Mendeleyev (1834–1907) uncovered this organizing principle of the table. He noticed that when the elements are arranged (more or less) by atomic weight, certain properties – the aforementioned nuclear and chemical properties – repeat periodically across the resulting table or chart. Thus, the name of the "Periodic" Table.

In fact, the guiding principle behind the periodic nature of the table is atomic number (the number of protons in the nucleus), not weight. In addition, much of the information needed to construct an accurate table was either missing or incorrect. Mendeleyev's audacious gambit was to hazard that, where the existing evidence did not match his scheme, he was right and science had, hitherto, been wrong.

Elements in the periodic table are sorted into seven rows or periods, with the atomic number increasing as you move along the row from left to

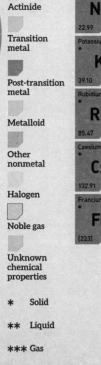

Legend:
- Alkali metal
- Alkali earth metal
- Lanthanide
- Actinide
- Transition metal
- Post-transition metal
- Metalloid
- Other nonmetal
- Halogen
- Noble gas
- Unknown chemical properties

* Solid
** Liquid
*** Gas
**** Unknown

Hydrogen *** H 1.008 1							
Lithium * Li 6.941 3	Beryllium * Be 9.012 4						
Sodium * Na 22.99 11	Magnesium * Mg 24.31 12						
Potassium * K 39.10 19	Calcium * Ca 40.08 20	Scandium * Sc 44.96 21	Titanium * Ti 47.87 22	Vanadium * V 50.94 23	Chromium * Cr 52.00 24	Manganese * Mn 54.94 25	Iron * Fe 55.84
Rubidium * Rb 85.47 37	Strontium * Sr 87.62 38	Yttrium * Y 88.91 39	Zirconium * Zr 91.22 40	Niobium * Nb 92.91 41	Molybdenum * Mo 95.94 42	Technetium * Tc [98] 43	Ruthenium * Ru 101.07
Caesium * Cs 132.91 55	Barium * Ba 137.33 56	LANTHANIDES ▼	Hafnium * Hf 178.49 72	Tantalum * Ta 180.95 93	Tungsten * W 183.84 96	Rhenium * Re 186.21 75	Osmium * Os 190.23
Francium * Fr [223] 87	Radium * Ra [226] 88	ACTINIDES ▼	Rutherfordium **** Rf [267] 104	Dubnium **** Db [268] 105	Seaborgium **** Sg [269] 106	Bohrium **** Bh [270] 107	Hassium **** Hs [269]

Lanthanum * La 138.91 57	Cerium * Ce 140.12 58	Praseodymium * Pr 140.91 59	Neodymium * Nd 144.24 60	Promethium * Pm [145] 61	Samarium * Sm 150.36
Actinium * Ac [227] 89	Thorium * Th 232.04 90	Protactinium * Pa 231.04 91	Uranium * U 238.03 92	Neptunium * Np [237] 93	Plutonium * Pu [244]

right. This arranges the elements into columns that are called families because, in each column, the elements share a "family resemblance", with similarities in physical and chemical properties. This is just one of many different ways to divide up the table. Another is into metals (conductive, usually solid elements that tend to be shiny, ductile and malleable), non-metals (non-conductive gases, solids and liquids) and metalloids (semi-metals, often semi-conductive and, therefore, useful in electronics).

All of life is contained within the mysteries of the periodic table.

What supposedly iconic mystery of life is specifically represented by the following two sequences of elements?

- Cr – Hf – Re – W – Mo – Sg
- Pd – Cu – Zn – In – Ds – Nh

					Helium *** He 4.003 2
Boron * B 10.81 5	Carbon * C 12.01 6	Nitrogen *** N 14.01 7	Oxygen *** O 16.00 8	Fluorine *** F 19.00 9	Neon *** Ne 20.18 10
Aluminium * Al 26.98 13	Silicon * Si 28.09 14	Phosphorus * P 30.97 15	Sulfur * S 32.07 16	Chlorine *** Cl 35.45 17	Argon *** Ar 39.95 18

Cobalt Co 3 27	Nickel * Ni 58.69 28	Copper * Cu 63.55 29	Zinc * Zn 65.39 30	Gallium * Ga 69.72 31	Germanium * Ge 72.63 32	Arsenic * As 74.92 33	Selenium * Se 78.96 34	Bromine ** Br 79.90 35	Krypton *** Kr 83.80 36
Rhodium Rh 91 45	Palladium * Pd 106.42 46	Silver * Ag 107.87 47	Cadmium * Cd 112.41 48	Indium * In 114.82 49	Tin * Sn 118.71 50	Antimony * Sb 121.76 51	Tellurium * Te 127.60 52	Iodine * I 126.90 53	Xenon *** Xe 131.29 54
Iridium Ir 22 77	Platinum * Pt 195.08 78	Gold * Au 196.97 79	Mercury ** Hg 200.59 80	Thallium * Tl 204.38 81	Lead * Pb 207.2 82	Bismuth * Bi 208.98 83	Polonium * Po [209] 84	Astatine * At [210] 85	Radon *** Rn [222] 86
Meitnerium Mt [8] 109	Darmstadtium **** Ds [281] 110	Roentgenium **** Rg [281] 111	Copernicium **** Cn [285] 112	Nihonium **** Nh [286] 113	Flerovium **** Fl [289] 114	Ununpentium **** Uup [289] 115	Livermorium **** Lv [293] 116	Ununseptium **** Uus [294] 117	Ununoctium **** Uuo [294] 118

Europium Eu 96 63	Gadolinium * Gd 157.25 64	Terbium * Tb 158.93 65	Dysprosium * Dy 162.50 66	Holmium * Ho 164.93 67	Erbium * Er 167.26 68	Thulium * Tm 168.93 69	Ytterbium * Yb 173.04 70	Lutetium * Lu 174.97 71
Americium Am [8] 95	Curium * Cm [247] 96	Berkelium * Bk [247] 97	Californium * Cf [251] 98	Einsteinium * Es [252] 99	Fermium * Fm [257] 100	Mendelevium * Md [258] 101	Nobelium * No [259] 102	Lawrencium * Lr [262] 103

RADIOACTIVITY

Nuclear chemistry is the branch of chemistry that deals with radioactivity, isotopes and nuclear reactions. As discussed, elements are distinguished from one another by the number of protons in their nuclei (i.e. their atomic number), which, in turn, has a bearing on the number of electrons normally orbiting the nucleus. This, in turn, determines the chemical properties of the element.

However, protons are not the only subatomic particles in the nucleus; there can also be neutrons, and we have already seen that the number of these can vary in different isotopes of the same element. In general, as nuclei get bigger, they become less stable. Sometimes they spontaneously break up, either by ejecting particles and/or electromagnetic waves; this is collectively known as radioactivity. The nuclei can also be split into two smaller ones, which is known as nuclear fission, and releases very large amounts of energy – for example in nuclear bombs.

PUZZLE INSTRUCTIONS

"Decompose" each row and column into separate sections. Shade two squares in every row and column, so that the number of unshaded squares between those two squares is equal to the number at the start of the corresponding row or column. Shaded squares do not touch, not even diagonally.

	5	1	5	1	7	3	3	1	1
1									
1									
2									
1									
1									
1									
1									
1									
2									

HOW MANY BARRELS OF PISS DOES IT TAKE TO DISCOVER A NEW ELEMENT?

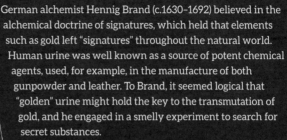

German alchemist Hennig Brand (c.1630–1692) believed in the alchemical doctrine of signatures, which held that elements such as gold left "signatures" throughout the natural world. Human urine was well known as a source of potent chemical agents, used, for example, in the manufacture of both gunpowder and leather. To Brand, it seemed logical that "golden" urine might hold the key to the transmutation of gold, and he engaged in a smelly experiment to search for secret substances.

Brand let 60 tubs of urine putrefy in his basement, and then boiled it down to a paste. Through heating the paste with charcoal, he was able to extract not gold, but a mysterious waxy substance that glowed in the dark. He named this new element phosphorus, after the Greek for "bringer of light".

PUZZLE INSTRUCTIONS

Divide the 60 tubs, represented by squares in the below grid, into different elements with different numbers of particles. Fill each empty cell with a number such that every number in the grid is part of a continuous region of that many cells. A region is continuous wherever two cells of the same value touch. Two different regions made up of the same number of cells cannot touch. You may find it helpful to draw borders to indicate the regions.

	11	5	6		6		
				6		7	
		5		6	6	7	
		5	2				4
11				8	3		
	3	8	6		6		
	5		5				
		5		3	3	3	

AN EXPLOSIVE DISCOVERY

Since the time of the legendary first emperor, Qin Shi Huang, Chinese alchemists had sought to combine substances to produce magical effects, but never with the explosive results experienced by some unfortunate ninth-century CE experimenters. An alchemical text from around the year 850CE records what happened when alchemists mixed charcoal from burned willow with honey, sulphur and saltpetre (potassium nitrate, obtained in ancient times from urine or bird droppings):

"Some have heated together the saltpetre, sulphur and carbon of charcoal with honey; smoke and flames result, so that their hands and faces have been burnt; and even the whole house burnt down."

The charred alchemists had inadvertently created gunpowder, a mixture in which saltpetre acts as a powerful oxidizer, powering the carbon to oxidize rapidly and explosively into carbon dioxide, with the sulphur acting to catalyse the reaction. Although initially used as a pesticide and skin treatment, it later became a key military technology, changing the course of warfare down the centuries with immense social and economic ramifications.

				3		2	
3		4			4		2
1		1		3			3
1			1				
2					4		3
	5	4			2		1
3			3	3		3	
		4				2	

PUZZLE INSTRUCTIONS

Locate the explosives! Place mines into some empty cells in the grid. Clues in some cells show the number of mines in touching cells – including diagonally. No more than one mine may be placed per cell.

WHAT IS A CHEMIST'S FAVOURITE ANIMAL?

Chemists love moles, although their moles are rather different from the gardener's bane. In chemical jargon, a mole is the amount of a substance that contains Avogadro's number of particles. Avogadro's number, or constant, is defined as the number of atoms in exactly 12g of carbon-12, and has been experimentally determined as 6.0221367 $\times 10^{23}$, or roughly *600 billion trillion*.

Named after the Italian lawyer and scientist Amadeo Avogadro, the Avogadro number is a vital constant for chemistry, as it allows chemists to relate atomic mass to measurable mass in the real world, via the concept of the mole. The mole is therefore an immensely powerful concept because it provides a simple way to measure – in relative terms – how many atoms or molecules of a substance are present. In the real world, it is effectively impossible to count atoms or molecules but, with the mole, you can at least count by weighing.

The definition of the mole means that an Avogadro's number of particles weighs the same in grams as the atomic weight of the particle. Thus, a mole of carbon-12 weighs 12g. If you have 48g of carbon-12, you know that this is equal to 4 moles.

How is this helpful in practice? Well, imagine you find that this amount of carbon reacts with 128g of O_2 gas (also 4 moles). You can then deduce that the resulting gas is carbon dioxide, not carbon monoxide, a vital distinction!

▼ Not the right type of mole, but it's quite possible some chemists are partial to this one as well.

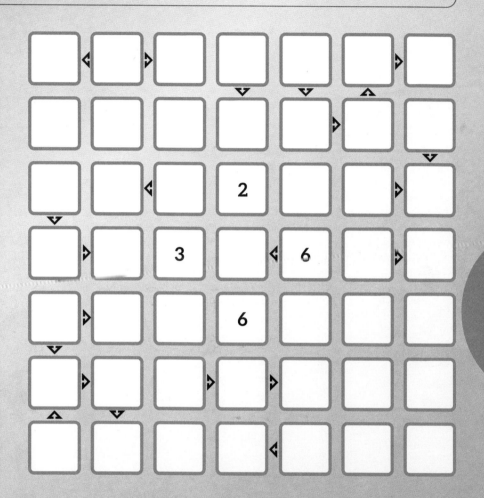

PUZZLE INSTRUCTIONS

Calculate the weight of each square, based on the inequalities shown. Place 1 to 7 once each into every row and column while obeying the inequality signs. Greater than (>) signs between some cells indicate that the value in one cell is greater than that in another, as indicated by the sign. The sign always points towards the smaller number.

BUCKYBALLS AND NANOTUBES

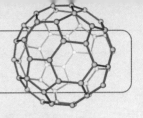

CHEMISTRY

Carbon atoms can form up to four bonds with other atoms. This gives them fascinating flexibility to form different molecules, including the long-chain, complex molecules that make up living things. They can also combine exclusively with other carbon atoms to form a number of different allotropes (pure forms) of carbon. Some of these are familiar, such as charcoal, or the graphite used in pencil leads, or even diamonds. But some, such as buckyballs and nanotubes, are exotic.

A buckyball is the informal name for a molecule of buckminsterfullerene, a soccer ball shaped sphere of 60 carbon atoms named after the architect and futurist Buckminster Fuller. The nickname arose because of the close resemblance in the structure of a hemisphere of buckminsterfullerene to Fuller's architectural invention, the geodesic dome. It was one of the first nanomolecules ever discovered.

Carbon molecules can also be arranged into nanotubes, which are rolled up sheets of graphene. This is an allotrope of carbon in which the atoms are linked only in a horizontal plane, to give a sheet just one atom thick. Nanotubes display amazing properties, such as high conductivity and enormous tensile strength, and many believe that they will be the key to many exciting emerging nanotechnologies. They may, for example, one day replace silicon in our computer chips, allowing for incredible advances in computing performance.

PUZZLE INSTRUCTIONS

Connect the atomic particles to create a complex molecule. Join circled numbers with horizontal or vertical lines. Each number must have as many lines connected to it as specified by its value. No more than two lines may join any pair of numbers, and no lines may cross. The finished layout must connect all numbers, so you can travel between any pair of numbers by following one or more lines.

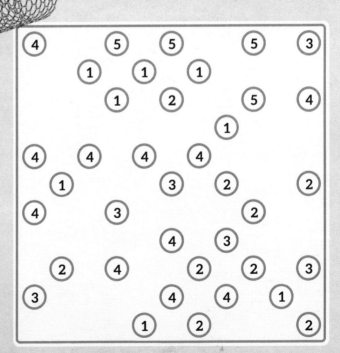

WATER IS WEIRD

Water, the liquid phase of the molecule dihydrogen oxide, is perhaps the most important substance on Earth, but also one of the strangest in terms of its chemistry. Its unusual chemistry is a consequence of the shape of the molecule: the two hydrogen atoms do not attach on either side of the central oxygen atom (i.e. with an angle of 180° between the bonds), but at angle of just 105°. This effectively gives the molecule "ends", which, in turn, means that charge is not distributed evenly across it.

The slightly charged ends weakly attract one another, forming easily broken "hydrogen bonds". These bonds make water molecules much stickier than other molecules of similar size and composition. As a result, water has a much higher boiling point than would otherwise be expected, so water remains liquid across a much wider range of temperatures. It can also absorb a lot of heat and only releases it slowly, thus helping to regulate and buffer temperature changes on Earth, whereas other planets like Mars experience wild swings of temperature. The slightly charged nature of the molecule also makes water a brilliant solvent, and this, in turn, makes possible most of the chemistry of life.

PUZZLE INSTRUCTIONS

Can you work out how the water flows? Shade some squares to represent water, so that each row and column contains the given number of shaded squares. Place the water so that all connected areas have the same highest water level.

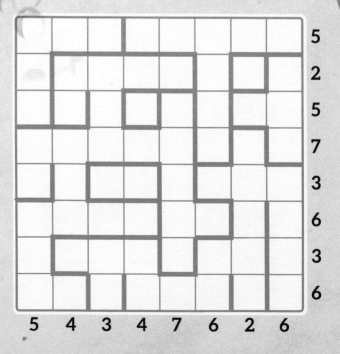

Row clues (top to bottom): 5, 2, 5, 7, 3, 6, 3, 6

Column clues (left to right): 5, 4, 3, 4, 7, 6, 2, 6

HOW TO BUILD A BETTER BATTERY

A battery is a device for producing a steady supply of direct electric current through the conversion of chemical energy to electricity. Many key modern technologies, such as mobile phones or electric cars, depend on lithium ion batteries. In these, ions of the metal lithium sit in a liquid medium (called the electrolyte) and travel back and forth between the positive and negative electrodes, carrying electrical charge.

Because the process is reversible in a lithium ion battery, it can be recharged, which is a decided advantage. However, the energy density of lithium ion batteries – the amount of energy they can hold per unit of mass – is the primary limiting factor for all manner of battery-powered technologies. For example, if we want to have long-range electric cars, airplanes, boats and drones, or useful autonomous robots, we will need batteries with much higher energy densities.

A number of strategies have been tried to achieve this, including replacing the substances used for the electrodes with lighter materials, such as sulphur or silicon; changing liquid electrolytes to solids; effectively making the electrodes into liquids (known as a flow battery); or doing away with batteries altogether and turning the whole structure of a vehicle into a capacitor (a device that stores electrostatic charge).

This is one of the primary challenges facing science over the next couple of decades, and it will be especially important in the fight against climate change.

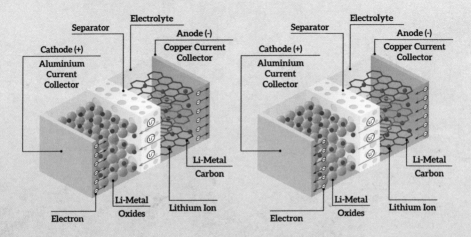

DISCHARGE

CHARGE

Separator
Electrolyte
Cathode (+)
Anode (-)
Copper Current Collector
Aluminium Current Collector
Li-Metal Carbon
Li-Metal Oxides
Lithium Ion
Electron

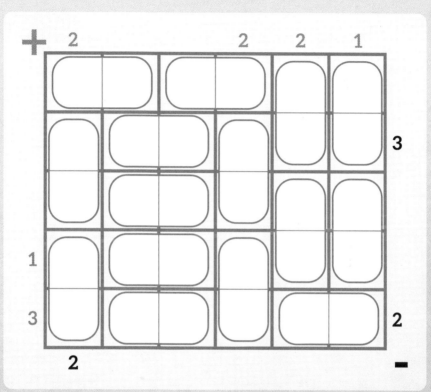

PUZZLE INSTRUCTIONS

Place the batteries to avoid like-poles touching. Locate the sets of batteries. Each oval must either be shaded to mark it as empty, or marked with an anode (+) at one end and cathode (-) at the other. Same terminals cannot touch to the left/right/above/below. Some numbers outside the grid reveal the total number of anodes (on the top and to the left of the grid) and cathodes (on the bottom and to the right of the grid) in each row and column, but not all numbers are given.

KEKULÉ'S DREAM

Thanks to its ability to form lots of bonds – and to link together in long chains that can act as skeletons to which other elements can bond – carbon can produce an almost limitless array of molecules. This versatility helps explain why carbon is the basic element of life.

The study of carbon molecules is called organic chemistry. Organic chemicals are generally made from very few elements: carbon and hydrogen, sometimes oxygen and nitrogen, and occasionally a handful of others. However, these apparently simple ingredients disguise an immense complexity, and early organic chemists struggled to understand how even quite small organic compounds were structured. For instance, pioneering organic chemist Friedrich August Kekulé (1829–96) was struggling to work out the structure of benzene (a relatively simple molecule made of just six carbon atoms and six hydrogen atoms), until the answer famously came to him in a dream.

"I turned my chair to the fire and dozed. Again, the atoms were gambolling before my eyes... all twining and twisting in snakelike motion. But look! What was that? One of the snakes had seized hold of its own tail, and the form whirled mockingly before my eyes. As if by a flash of lightning I awoke..."

Benzene, he now understood, had a cyclic or ring-shaped structure.

PUZZLE INSTRUCTIONS

Create a benzene-like loop, tracing the path of the delocalised electron. Draw a single loop through some empty cells, made up of horizontal and vertical lines between cell centres. The loop does not cross or overlap itself, and can only pass through empty grid cells. Cells with numbers indicate how many touching cells the loop passes through, including diagonally touching cells.

3							
		7				8	
	7						
							3
				4			3
3			6		5		
			4	4			

FINGERPRINTS OF LIGHT

In the early nineteenth century, German optician Joseph von Fraunhofer (1787–1826) looked at a flame through an optical glass, which refracted the light to spread out the different colours along a spectrum. He noted that there were some distinct lines of brightness: the flame was emitting some wavelengths at greater intensity than others. Fraunhofer turned his glass to the sun and discovered that the otherwise continuous spectrum was actually broken up by a number of dark lines.

He labelled them with letters, but their origin and meaning remained a mystery for decades. It wasn't until 1859 when Robert Bunsen (1811–99), a professor of chemistry at Heidelberg who was curious about the distinctive colours obtained by burning certain elements, recruited his colleague from the physics department, Gustav Kirchhoff (1824–87), to invent spectral analysis. They showed that the light emitted from burning elements displayed characteristic spectral patterns or fingerprints. They determined that by studying these patterns – which is now known as spectroscopy – they could identify the presence of elements within different materials.

Bunsen and others used the method to discover new elements, and he and Kirchoff were finally able to deduce that Fraunhofer's mysterious D-lines exist because sodium is present in the sun's atmosphere, absorbing certain wavelengths of light.

PUZZLE INSTRUCTIONS

Work out where to place the lights. Place light bulbs in some empty cells so that each empty cell is illuminated. Light bulbs illuminate every cell in their row and column until blocked by a black cell. No light bulb can be illuminated by another light bulb. Some black cells have numbers, which correspond to the number of light bulbs in their adjacent, non-diagonally touching, cells. Not all light bulbs are necessarily clued. All black cells are given.

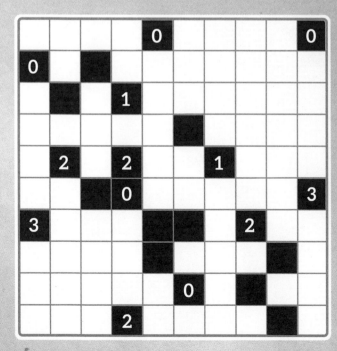

THE DEADLIEST POISON

Botulinum toxin is the most powerful poison known to humankind. Just a few hundred grams would be enough to kill everybody on the planet! Yet it is most famous today as a cosmetic treatment (Botox). It is produced by a bacterium, *Clostridium botulinum*, which is ubiquitous in the soil, but only threatens human health in cases such as contaminated tinned or bottled food or in poorly dressed wounds.

It is actually a mix of toxic proteins, which attack the junctions between nerve cells, preventing the release of neurotransmitters and thus stopping neurons from firing. Among other effects, this paralyzes muscles and can cause death by respiratory paralysis.

The name botulinum comes from the Latin *botulus*, meaning "sausage", after the first recorded outbreak, in Wildbad, Germany, in 1793, was traced back to contaminated blood sausage. Despite its dangers, botulinum can now be used in medicine and cosmetic surgery to treat conditions ranging from writer's cramp and squints to cerebral palsy and facial wrinkles.

PUZZLE INSTRUCTIONS

Reveal if there is poison by solving this hanjie puzzle: Shade some cells by obeying the clue constraints at the start of each row or column. The clues provide, in reading order, the length of every run of consecutive shaded cells in each row and column. There must be a gap of at least one empty cell between each run of shaded cells in the same row or column.

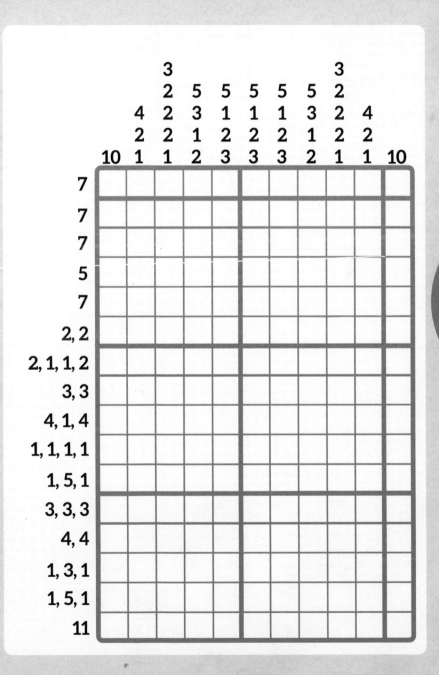

THE CHEMISTRY QUIZ

1) What is the chemical formula of water?

2) What do you get if you burn hydrogen in air?

3) With which element did the alchemists associate the sun and urine?

4) Which compound of two elements gives rust and blood their red colour?

5) What element is the lead in a pencil made from? Clue: it's not lead.

6) If you measured the pH level of a liquid and found that it was 1, would it be an acid or an alkali?

7) What is the smallest and lightest element in the periodic table?

8) What is the name given to an atom that has lost or gained electrons and so has an electric charge?

9) What is the densest element?

a) lead ☐

b) gold ☐

c) osmium ☐

10) The atomic number of oxygen is 8. How many protons does an atom of oxygen have?

11) What is the collective name for the group of elements that includes neon, xenon, krypton and argon?

12) Which metal can dissolve gold?

13) What is the name given to a cloud, or soup, of ions?

14) What elements make up a molecule of ethanol?

15) Saltpetre is one of the key ingredients of gunpowder. What is the chemical name of saltpetre?

a) potassium nitrate ☐

b) magnesium oxide ☐

c) nitrous oxide ☐

16) Which French scientist was the first to discover evidence of radioactivity?

a) Marie Curie ☐

b) Henri Becquerel ☐

c) Louis Pasteur ☐

17) The Haber Process is fundamental to the production of synthetic fertilizer. It combines nitrogen with hydrogen to produce which chemical?

18) What is the weight of a mole of carbon-12?

19) Which isotope of uranium is fissile (i.e. can be used to engender nuclear fission)?

a) U-238 ☐

b) U-235 ☐

c) U-232 ☐

20) Sodium has an atomic number of 11. What comes after it in the periodic table?

a) calcium ☐

b) potassium ☐

c) magnesium ☐

THE
ENVIRONMENT

There is no greater threat to the long-term fortunes of humanity than our own mistreatment of the environment. With both direct dangers – such as extreme weather and increased temperatures caused by greenhouse gases – and indirect issues – such as pandemics caused by the reckless encroachment of mankind into certain animals' natural habitats – looming, we need to change our habits, and swiftly. Fortunately, science can lead the way to a cleaner, greener future.

THE GREENHOUSE EFFECT

One of the reasons that Earth is hospitable to life, when similar planets such as Mars are not, is that our atmosphere includes high levels of what we call "greenhouse gases". These are gases such as carbon dioxide and water vapour, and they are named as such because they act on the Earth in the same way as glass panels of a greenhouse.

They are mostly transparent to short-wave radiation at the visible-light end of the spectrum, and so allow solar radiation through to warm the planet. However, they absorb long-wave radiation at the infra-red end of the spectrum, so when the warmed planet re-radiates its heat as infra-red, the energy cannot escape the atmosphere. Life on Earth is only possible because the greenhouse effect keeps the planet's temperature around 35°C higher than it would otherwise be without it.

However, as has become apparent in the past few decades, this effect can damage the Earth's fragile ecosystem as well. The burning of dirty fossil fuels, deforestation and various other human activities have increased the concentration of greenhouse gases in the atmosphere, and thus increased the intensity of this effect. As the Earth's temperature rises further, life will have to find new ways to survive in the changed circumstances.

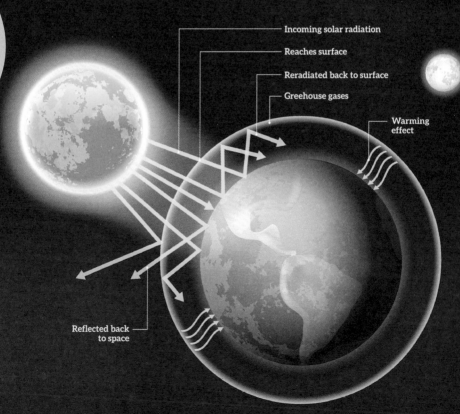

Incoming solar radiation

Reaches surface

Reradiated back to surface

Greehouse gases

Warming effect

Reflected back to space

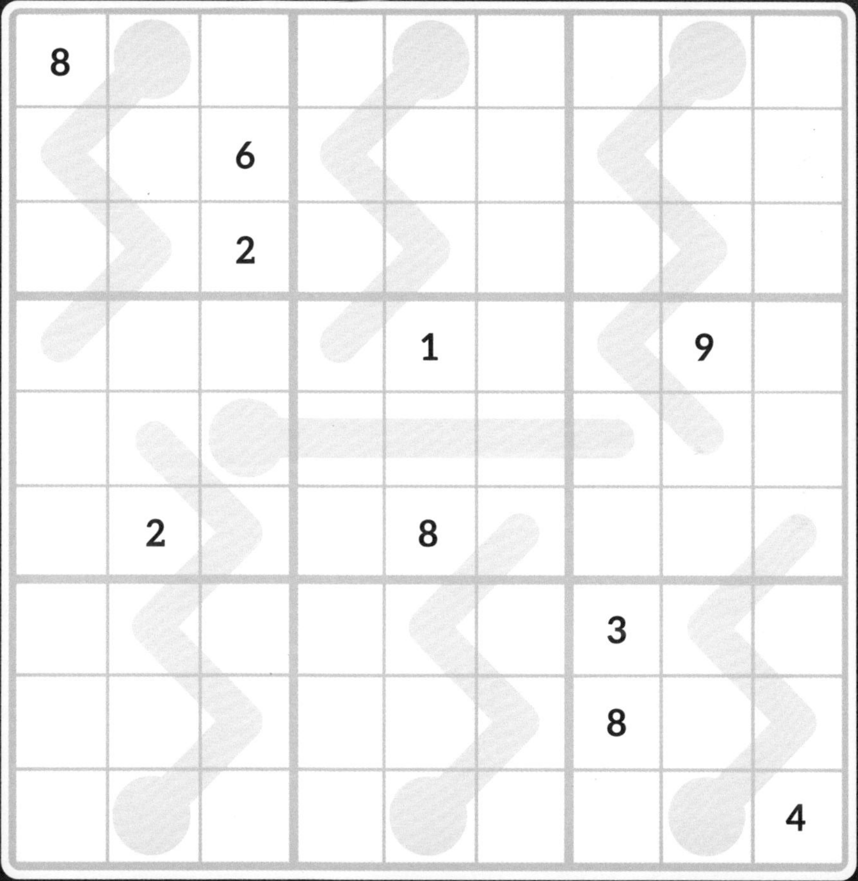

8								
		6						
		2						
				1			9	
	2			8				
						3		
						8		
								4

PUZZLE INSTRUCTIONS

Temperatures are rising! Place 1-9 once each into every row, column and bold-lined 3×3 box. The value of the digits along each shaded thermometer must increase cell by cell from the bulb (lowest value) to the head (highest value). This also means that digits cannot be repeated in a thermometer.

GEOENGINEERING

Geoengineering is the term used to describe planetary-scale interventions in global climate and ecology, currently often with the intent of ameliorating or preventing global heating, and thus climate change. Many geoengineering proposals sound like science fiction, such as a plan to block out a few per cent of the light reaching the Earth by sending "space parasols" to a point between the sun and the Earth.

Earthbound plans include proposals to expose vast regions of carbon-dioxide absorbing rock; seeding the oceans with masses of fertilizer to encourage algal growth that will draw carbon from the atmosphere; or triggering massive cloud formation to block incoming light. These all sound incredibly exciting, and the imagination of climate scientists appears to be increasingly fertile. However, beyond the obvious problems of cost and feasibility, geoengineering ideas raise the spectre of unintended consequences, as it is almost impossible to predict all the effects of such enormous interventions.

PUZZLE INSTRUCTIONS

Build a series of structures of varying heights to block out the sunlight. Place 1 to 5 once each into every row and column of the grid. Place these digits in the grid in such a way that each given clue number outside the grid represents the number of digits that are 'visible' from that point, looking along that clue's row or column. A digit is visible unless there is a higher digit preceding it, reading in order along that row or column. For example, in '21435' the 2, 4 and 5 are visible, but 1 is obscured by the 2, and 3 by the 4.

▼ A fantastical illustration of a potential space parasol.

TIPPING POINTS

The slow, steady creep upwards of global temperature, as a result of global heating, is cause enough for concern. But what really keeps climate scientists up at night is the prospect that the complex feedback systems of the global ecosystem will be – or already have been – pushed over crucial thresholds by climate change. These thresholds are known as tipping points because they will act to accelerate climate change, and thus might form feedback loops or vicious cycles that can't be reversed.

Examples include the loss of major rainforests, which currently act as massive carbon sinks, but which could burn down or be deforested, releasing masses of carbon and unlocking carbon stored in their soils; release of vast stores of methane (a greenhouse gas far more potent than CO2) currently locked in permafrost soils or submarine silt; and loss of reflective ice cover, which currently causes lots of incoming solar radiation to bounce back into space, instead of being soaked up by water or land. Prehistoric climate records reveal periods in which the Earth's climate underwent massive changes in timescales as short as a decade – far too quickly for civilization to adapt and mitigate – probably precisely because of such tipping points.

▼ Snow-capped mountains in Antarctica will hopefully not become a thing of the distant past.

7	5	1
1	8	1
8	2	8

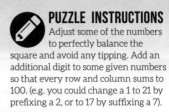

PUZZLE INSTRUCTIONS

Adjust some of the numbers to perfectly balance the square and avoid any tipping. Add an additional digit to some given numbers so that every row and column sums to 100. (e.g. you could change a 1 to 21 by prefixing a 2, or to 17 by suffixing a 7).

ICE AGES

Early nineteenth-century European geologists were confronted with a lot of unexpected evidence for relatively recent reshaping of many landscapes on a colossal scale. Thick beds of clay and gravel; long parallel scars in naked rock; huge erratics (which are large boulders deposited far distant from matching rock formations); ancient beaches far above sea level. Traditionally, these were held to be evidence of the Biblical flood, or something similar, but this was not enough for inquisitive scientists.

The Swiss botanist and geologist Louis Agassiz argued that ice was responsible. Just a few tens of thousands of years before, he said, most of Northern Europe had been covered with a vast ice sheet. Glaciers had scoured the rock and carried boulders great distances, and when they melted, they deposited layers of clay and gravel. Meanwhile, landmasses rebounded once the weight of the ice was relieved, raising coastlines.

It is now known that the Earth has experienced multiple ice ages, although they rarely strike both the northern and southern hemispheres at the same time.

▼ A view of Franz Josef Glacier in New Zealand.

PUZZLE INSTRUCTIONS

Place the lone boulders in each marked glacial valley. Place boulders in some cells so that each row, column and outlined region contains exactly two boulders. These erratic boulders cannot touch each other – not even diagonally.

RENEWABLE ENERGY

Renewable energy is energy derived from sources that are continually replenished, rather than from finite resources such as coal, oil or natural gas. Almost all renewable types trace their energy source back to the sun. Water and wind power, for example, derive from cycles that are driven by solar energy (such as evaporation of water, which raises its height, giving it potential energy that can be converted into mechanical energy via waterwheels or turbines).

Some of the few exceptions include geothermal energy, which ultimately derives from the decay of radioactive elements in the Earth, and tidal energy, which derives from the influence of lunar gravity. Humanity will need to explore all of the possibilities that the various renewable energy pathways afford if we are to effectively combat the dangers of climate change.

PUZZLE INSTRUCTIONS

Fill in this grid, which wraps around in one constantly replenished continuous cycle. Place 1-9 once each into every row, column and bold-lined jigsaw shape. Some bold-lined jigsaw shapes "wrap around" the edges of the puzzle, continuing in the cell directly opposite. So, for example, in this puzzle the jigsaw region covering the first square in the second row continues at the opposite end of the same row.

JOURNEY TO THE CENTRE OF THE EARTH

The Earth can be imagined like a Scotch egg, a savoury snack comprising a hard-boiled egg wrapped in sausage meat, coated in breadcrumbs and deep-fried. The thin crust of the Scotch egg is like the outermost layer of the Earth. Below each is an intermediate layer that extends about halfway to the centre. In a Scotch egg, this layer is sausage meat; in the Earth it is partially molten rock known as the mantle.

A Scotch egg has a core composed of two layers: egg white surrounding egg yolk. The Earth, similarly, has a two-layer core, with molten iron and nickel around an inner core of solid iron and nickel. It's a long way from the surface to the centre. If you could drill a hole from the surface of the Earth to its centre, it would take about 19 minutes to fall to the bottom, assuming there was no air in the shaft to slow your fall.

▼ The Earth as a Scotch egg, with the thickness of each layer shown.

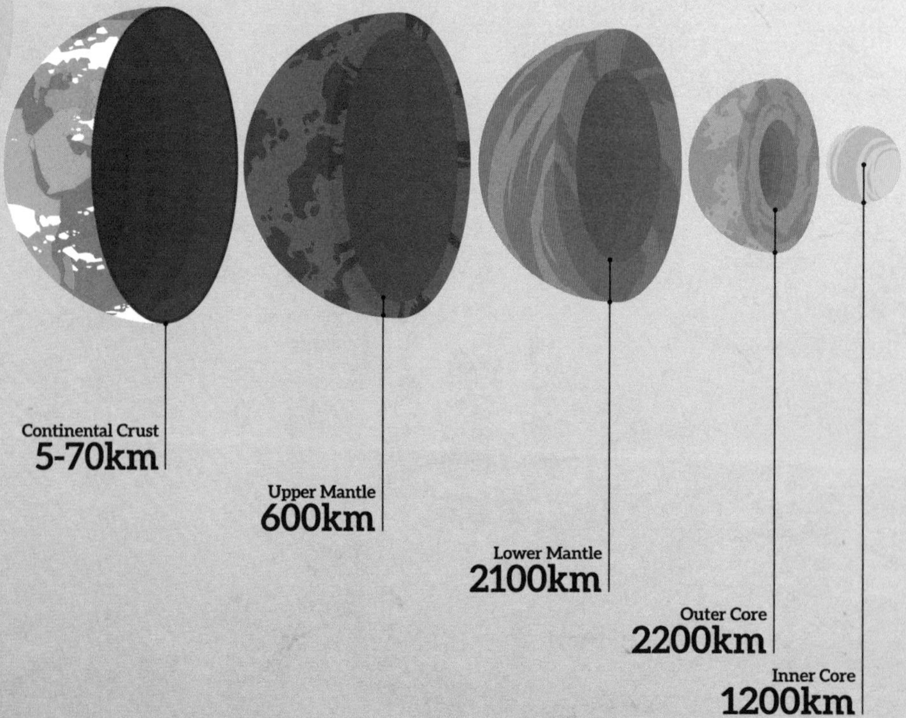

Continental Crust
5-70km

Upper Mantle
600km

Lower Mantle
2100km

Outer Core
2200km

Inner Core
1200km

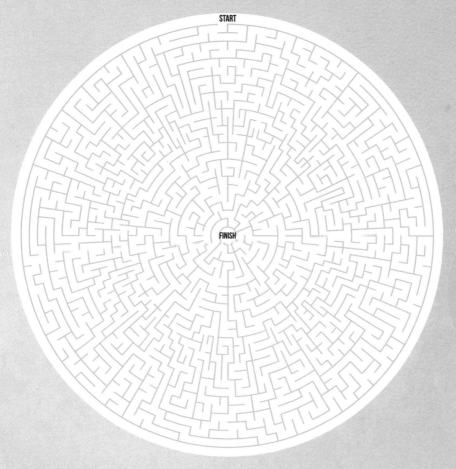

START

FINISH

PUZZLE INSTRUCTIONS
Journey to the centre of this Earth-
shaped maze. Find your way from the
entrance at the top to the exit in the centre.

HURRICANES

A hurricane is the name given to a severe tropical storm, but this is only the case in the North Atlantic and Caribbean. In the Pacific Ocean, such storms are called tropical cyclones. Whatever they are called, they are formed when warm air over the ocean picks up lots of water vapour. As the air rises, the vapour condenses into water drops, releasing energy, known as the latent heat of condensation.

This energy drives high winds, which then pick up more heat and vapour from the ocean. This feeds a positive feedback loop, which over time can build up to become enormously powerful and destructive. In a single day, a large hurricane can generate energy that's equivalent to 8,000 megatons (i.e. the power of 8 billion tons of TNT). This is greater than the destructive power of the world's entire nuclear arsenal!

PUZZLE INSTRUCTIONS

Can you work out the paths of the various different hurricanes, represented by the shapes in the grid? Draw a series of separate paths, each connecting a pair of identical hurricanes. No more than one line can enter any cell, and lines can only travel horizontally or vertically between cell centres.

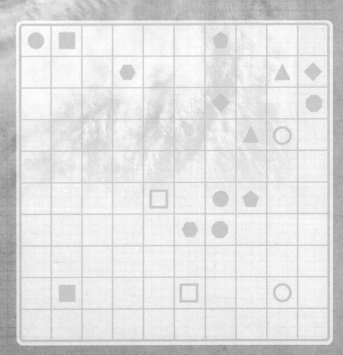

HOT AND WET

PUZZLE INSTRUCTIONS

Fill the thermometers according to the given clues. Shade the given number of squares in each row or column, as indicated by the row-/column-end numbers. No segment of a thermometer can be shaded unless all segments closer to the bulb, including the bulb itself, are already shaded.

▼ Participants in the 2018 Marathon des Sables making their way slowly though the desert.

Heatwaves can kill, and for humans even a relatively small change in body temperature – from around 37°C to 40°C – can be fatal. Yet we can also survive extraordinary temperatures and have been known to perform feats of physical endurance in extreme conditions. Each year, for example, some brave (or foolhardy) souls take part in the Marathon des Sables, a six-day ultramarathon held in the Sahara in temperatures of up to 50°C or more.

Events such as this are possible because our bodies can shed heat through evaporation of water. This occurs most obviously when we sweat, but is actually achieved mainly via breathing, with water evaporating from the lining of the lungs into air that we then exhale. As long as the inhaled air is dry, it doesn't matter how hot it is. Problems occur, however, if it is humid as well as hot. Then, we cannot lose heat by evaporation. Fortunately, at the moment, there is nowhere on the surface of the Earth where this happens. However, it is likely that global heating will cause such fatal heatwaves in the hottest regions like the Gulf of Arabia or along parts of the coasts surrounding the Indian Ocean within the next few decades. If this happens, anyone exposed to the hot, humid air will die of hyperthermia within around 20 minutes!

FALLING TO EARTH

In 2014, computer scientist Alan Eustace skydived from an altitude of 41.4 km, the highest parachute jump ever made. Eustace fell from near to the top of the stratosphere, from more than twice the height at which the airliner Concorde used to fly, and over four times higher than the peak of Mount Everest! However, he still wasn't even halfway to the edge of the atmosphere.

The Earth's atmosphere extends to roughly 100km up, where the Karman line is commonly held to signify the start of outer space. The topmost layer is called the thermosphere, and below it is the mesosphere. This extends down to the stratopause, where the stratosphere begins, extending down to the tropopause, the boundary of the troposphere, which is at about 12km altitude. It is in this lowest layer that 80 per cent of the mass of the atmosphere is found, and where almost all of our weather occurs.

▼ The very important ozone layer lies in the stratosphere, and the exobase marks the very edge of the Earth's atmosphere.

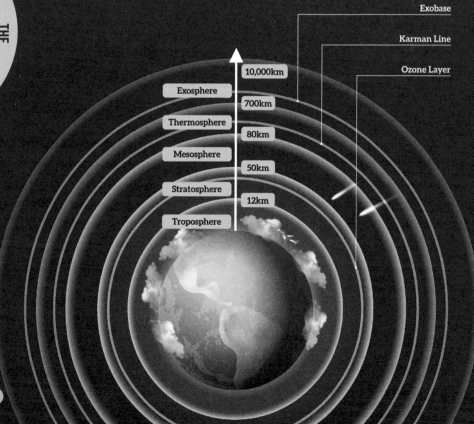

Exobase

Karman Line

Ozone Layer

Exosphere	
	10,000km
	700km
Thermosphere	
	80km
Mesosphere	
	50km
Stratosphere	
	12km
Troposphere	

PUZZLE INSTRUCTIONS

Can you reveal the relevant picture by solving this hanjie puzzle?
Shade some cells by obeying the clue constraints at the start of each
row or column. The clues provide, in reading order, the length of every run of
consecutive shaded cells in each row and column. There must be a gap of at least
one empty cell between each run of shaded cells in the same row or column.

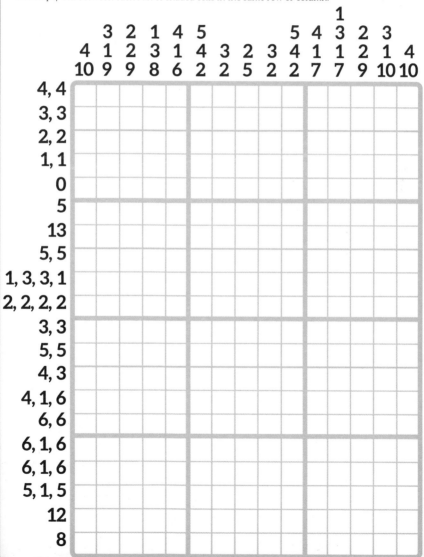

FIRE SEASON

Apocalyptic bush fires have, in recent years, ravaged places as far-flung as Australia, Siberia and California. These regions have experienced bush fires for millennia, and plants, animals and indigenous peoples have adapted to cope and, in some cases, to thrive. But the fires are getting worse, and more frequent – six of the worst ten fires in California's history, for instance, have occurred since 2017. Global heating is an obvious contributory factor, in more than just the most obvious way.

Not only are dry seasons becoming hotter and drier, but more intense wet periods – the result of more energy in the atmosphere – are causing rapid but short-lived growth of young, low-level bush and woodland, which soon dies off, dries out and forms the perfect feedstock for bush fires. These problems are exacerbated by human management of wilderness, which for decades has suppressed natural fires to protect encroaching settlements, while managing woodlands in the opposite way to how nature has traditionally managed them, extracting large, old trees while encouraging dense, younger growth.

If we are to break the cycle of worsening fires occurring each year, we are going to have to reimagine how we manage our forests and woodlands, while simultaneously addressing the long-term problems causing global heating.

PUZZLE INSTRUCTIONS

Create a firebreak around the fire before it spreads. Join all the dots to form a single firebreak loop. The loop cannot cross or touch itself at any point. Only horizontal and vertical lines between dots are allowed. Some parts of the loop are already given.

▼ A smoldering bushfire in the Australian Outback.

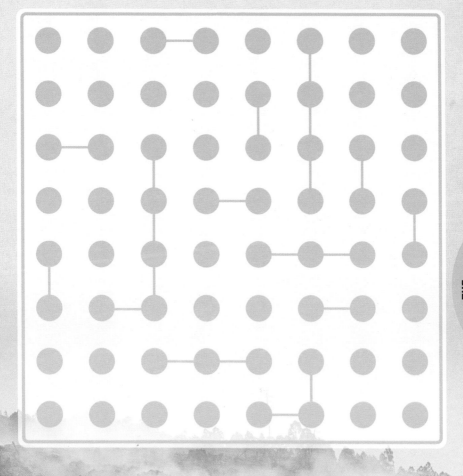

PREDICTING EARTHQUAKES

For millennia, scientists have attempted to study and understand earthquakes. The poet and engineer Zhang Heng (78–139 CE) – who can been seen as the Leonardo da Vinci of second-century CE China – is now most famous for his *Houfeng didong yi*, or "earthquake weathervane". It was a sophisticated seismoscope (an earthquake-sensing instrument) in which balls were released from a central urn to roll out of the mouth of a bronze dragon and into the mouth of a bronze toad, thus indicating the direction of distant quake epicentres.

Despite this long history, earthquake prediction has eluded scientists, probably because of the complex variables and random nature of some factors. A useful prediction needs to define the date and time of a quake, along with its location and its magnitude. Even the most advanced machine-learning algorithms are not able to provide useful levels of accuracy, and the very nature of earthquakes may mean that prediction is inherently impossible. As the US Geological Survey says, "We do not know how [to predict earthquakes], and we do not expect to know how any time in the foreseeable future."

PUZZLE INSTRUCTIONS

Work out the directions of each earthquake, according to their given magnitude. Draw one or more horizontal or vertical lines emanating from each numbered cell. Lines cannot cross numbered cells or enter the same cell as another line. Each number indicates how many cells its lines travel into; the numbered cells themselves are not counted. All cells must be visited by a line. Each line can only connect to a single number.

						8	
				5			
		6					
3				4			
			4				4
						6	
				5			
		9					

MEGA-TSUNAMIS

Tsunamis are giant waves caused by displacement of the land under or next to water, typically by earthquakes under the ocean. Ordinary tsunamis are terrifying and destructive enough, as the Boxing Day tsunami of 2004 and the Tohoku tsunami of 2011 testify, but coastlines in some parts of the world bear fossil witness to the occurrence of much bigger waves, known as mega-tsunami.

These are thought to result from massive landslides, either in shallow water or from mountainsides, which displace water in the same way that dropping a large rock at one end of a bathtub will cause water to slop over the other end. There are a number of potential instances of this that are currently worrying scientists. One possible flashpoint for such a disaster might be the collapse of the flank of the Cumbre Vieja volcano on La Palma in the Canary Islands. If this slid into the sea, experts believe it could create a tsunami 100 metres high in the immediate vicinity. The wave would travel across the Atlantic at the speed of a jumbo jet, and still be 10m high when it hits Brazil, while parts of the North American coastline could funnel the wave into a monster 50m high!

PUZZLE INSTRUCTIONS

In this puzzle, the effect of each number ripples out, like waves of varying sizes. Place a digit in each empty cell so that each bold-lined region contains each digit from 1 to the number of cells in that region. Equal digits in a row or column must be separated by at least the number of cells given by that digit. For example, a pair of '3's must have at least three other cells between them.

						1	
4		3		1			3
			1				
				1		1	
1	2				1		
		2				5	
	4		1				
			2			2	

THE NEXT SUPER-VOLCANO

In around 71,500 BCE, the volcano of Toba, in Northern Sumatra, exploded in the biggest volcanic eruption that humans have probably ever witnessed. Thousands of cubic kilometres of rock and ash were blasted into the atmosphere, along with billions of tonnes of sulphur, creating aerosols that blocked 99 per cent of sunlight. These plunged the Earth into a long-term volcanic winter.

The evidence that can be found in our own DNA shows that humanity was nearly wiped out by this event and that, for 20,000 years, there were only a few thousand humans left on the planet. Much larger volcanic events, known as flood basalts, have occurred in geological history, and they were probably at least partly responsible for mass extinctions that occurred 65 million and 250 million years ago.

Could such eruptions happen today? It is known that there are huge bubbles of magma not far beneath the surface of Yellowstone National Park in the US, and below the Phlegrean Fields region of southern Italy. Yellowstone has been the site of super-volcanic eruptions at intervals of roughly 650,000 years, and it has been 650,000 years since the last one...

▼ A stunning geyser in Yellowstone National Park.

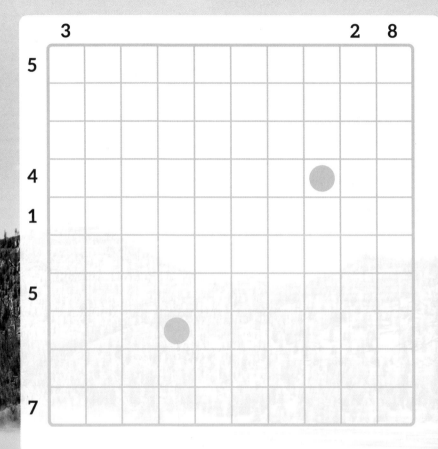

PUZZLE INSTRUCTIONS

Track the pyroclastic flow of this eruption. Shade some cells to form a single lava-snake that starts and ends at the given cells. A lava-snake is a path of adjacent cells that does not branch or cross over itself. The lava-snake does not touch itself – not even diagonally, except when turning a corner. Numbers outside the grid specify the number of cells in their row or column that contain part of the lava-snake.

THE ENVIRONMENT QUIZ

1) At which pole would you find Antarctica?

2) Which natural phenomenon is also known as a twister?

3) Which is the biggest continent?

4) What is the biggest river in the world by volume of discharge?

5) The Amazon rainforest spreads across nine countries, including Brazil, Suriname, Peru, Guyana and French Guiana. Can you name any of the other four countries?

6) What is the name given to the flow of water from the land to the rivers to the sea and then back again via evaporation, cloud formation and precipitation?

7) Name the five ocean basins recognized by the US Board on Geographic Names

8) In which layer of the atmosphere do humans live?

9) What source of energy is being tapped in a geothermal power plant?

10) In which direction does a hurricane spin?

11) What is the name of the layer of the Earth beneath the crust?

12) What is the lowest point on the Earth's surface?

13) Which is the most potent greenhouse gas?

a) carbon dioxide ☐

b) methane ☐

c) nitrous oxide ☐

14) An earthquake registering 6 on the Richter Scale is how many times more powerful than one registering 5?

15) What is the name given to the global cooling that can occur after a huge volcanic eruption?

16) What is the opposite of a neap tide?

17) What name is given to the periods between Ice Ages?

a) inter-pluvials ☐

b) warming intervals ☐

c) inter-glacials ☐

18) What is the name of the scale used to rate the impact hazard of asteroids and comets?

a) the Milano Scale ☐

b) the Torino Scale ☐

c) the Firenze Scale ☐

19) Roughly how many trees are there in the world?

a) 300 million ☐

b) 3 billion ☐

c) 3 trillion ☐

20) What is the name given to the climate cycle in the Pacific Ocean, which has a global impact on weather patterns, and which occurs every few years?

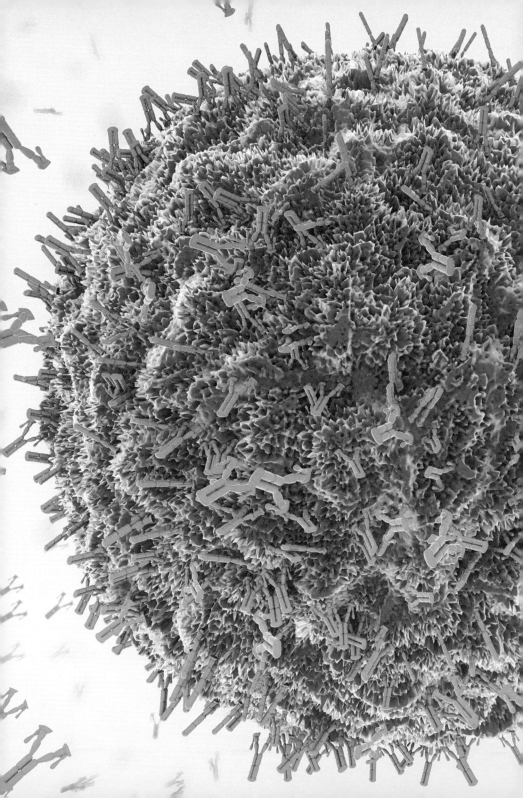

HEALTH

Germs, such as viruses and bacteria, were almost completely unknown until the 19th century. Such have been our advances in the last 150 years that – if you live in a country with a robust healthcare system – when you visit the hospital, you fully expect to come home healthy and with all of your limbs still intact. This has *not* been the case for most of human history. However, recent events have taught us just how much we must still learn to keep the world's population happy and healthy.

SYNTHETIC DNA

DNA is the genetic material that encodes the blueprint for every organism. In the 20th century, biologists learned to sequence DNA and therefore read the genomes (the full DNA sequence) of relatively simple organisms. We can now sequence the DNA of humans without too much trouble, but some scientists began to wonder just how simple a genome could be. If they could work out the minimum genome necessary to create a viable organism, perhaps they could write their own version: a new page in the book of life.

A team led by controversial American geneticist Craig Ventor has attempted to do just this, writing a simple but complete genome stitched together from synthetic DNA, before inserting it into a biological shell (a microbe from which the DNA had been "deleted", or removed). The resulting organism was dubbed JCVI-syn 3.0, or "Synthia", a living fully synthetic microbe

PUZZLE INSTRUCTIONS

Place the four different base letters of non-synthetic DNA – A (adenine), G (guanine), C (cytosine) and T (thymine) – into the grid. Place A, C, G and T once each into every row and column within the grid. This means that there will be one empty square in every row and column. Each letter given outside the grid must match the closest letter to it in within the same row/column. Letters may not share squares.

▶ An artistic visualization of the structure of DNA.

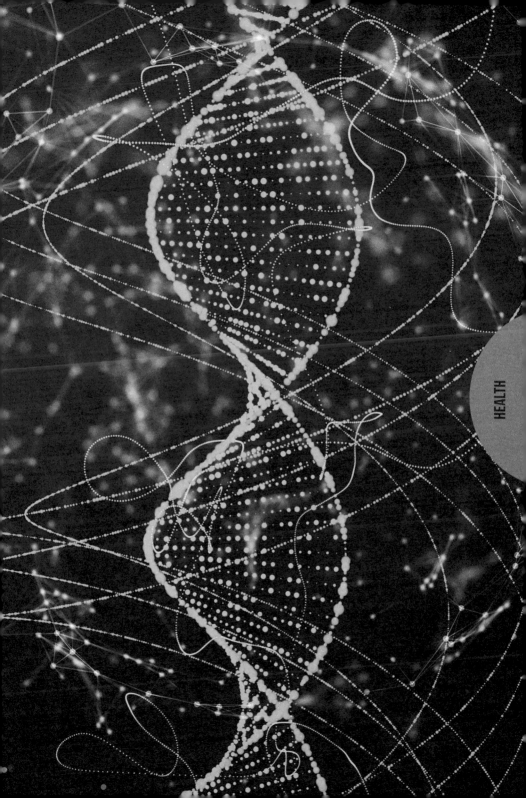

BIONIC PROSTHETICS

The well-known word "bionic" is actually a contraction of "biological electronics", which refers to the technology of interfacing electronics with biological systems. In medicine, the goal of bionics has been to create replica organs and limbs that can function as well as – or even better than – biological versions, and which can interact with the control systems of the body.

The ultimate aim is to be able to attach an electromechanical arm or leg to the body that moves in response to nerve signals from the user's brain, and which can send sensory information back to the brain about touch and position. Examples of current advanced bionic prosthetics include thought-controlled robotic arms, and cochlear and retinal implants.

▼ Currently in the realms of science fiction, but not for long!

PUZZLE INSTRUCTIONS

Match the "person" with the "prosthetic" that they need. Draw a series of separate paths, each connecting a pair of identical numbers. No more than one path can enter any square, and paths can only travel horizontally or vertically between squares.

HEALTH

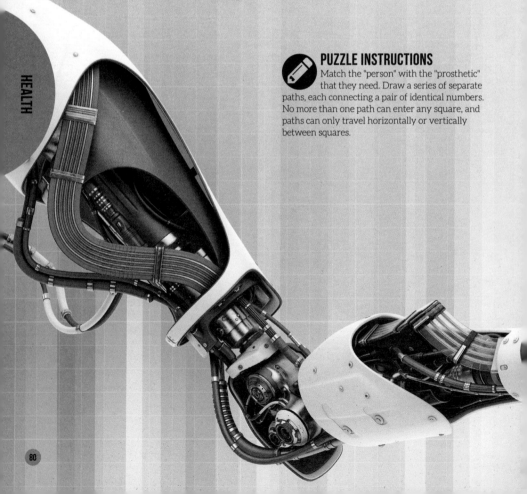

1 2 3 4

4

5 6 6

7 5 8

1 8 2

9

10 10

3

11 11

9 7

CRISPR GENE EDITING

Genetic engineers have been able to insert new stretches of DNA into existing genomes since the 1970s, but the process has until recently been difficult, inefficient and inaccurate. However, the discovery of a bacterial antiviral defence system known as CRISPR-Cas9 has provided a powerful new tool that promises to make genetic engineering much more effective. In case it comes up in a pub quiz, CRISPR stands for "clustered regularly interspaced short palindromic repeats" and Cas stands for "CRISPR-associated enzyme".

In the natural world, the bacteria use the Cas9 enzyme to attack the DNA of invading viruses. Recently, genetic engineers learned to reprogramme CRISPR-Cas9 to hit, with great precision and potency, their own targets. This allows DNA to be selectively edited, allowing for the potential treatment of genetic illnesses, for example. CRISPR technology may herald the beginning of a new era of effective genetic engineering.

PUZZLE INSTRUCTIONS

"Snip up" the single shape into four identically shaped parts Draw along some of the dashed grid lines in order to divide this shape up into four regions. The regions must all be identical, although they may be rotated (but not reflected) relative to one another.

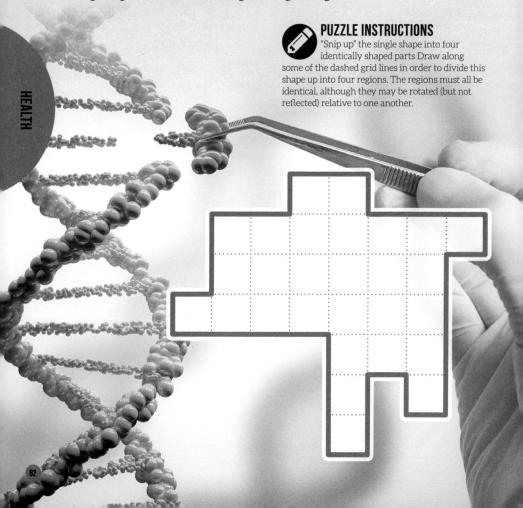

ANTIBODIES AND THE IMMUNE SYSTEM

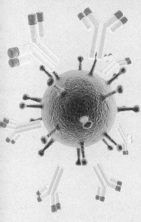

Your body is unfortunately something of a battleground. It is under constant attack from pathogens – organisms and substances that might harm it – but always fighting back with the powerful weapons of the immune system. These weapons range from passive barriers, such as your skin, to active "hunter-killer" agents, like white blood cells.

Such cells can use specialized proteins, known as antibodies, to identify, tag and latch on to pathogens. Antibodies are generally Y-shaped molecules, with targeting regions at the end of each arm. Specific antibodies can target specific molecules on the exteriors of pathogens – they are designed to perfectly fit those molecules – and your white blood cells manufacture an almost endless range of different antibodies, enabling them to recognize even pathogens you have never before encountered. Once an antibody makes contact with its target pathogen, a chain of processes is triggered, which includes ramping up production of that antibody throughout the immune system and "memorizing" it for rapid rollout in the future.

▲ A representation of Y-shaped antibodies attacking the "arms" of an invading virus.

HEALTH

PUZZLE INSTRUCTIONS

Split the grid into a set of rotationally symmetric phagocytes, each engulfing a bacteria in its exact centre. Draw along some of the grid lines in order to divide the grid up into a set of regions. Every region must contain exactly one circle, and the region must be symmetrical in such a way that if rotated 180 degrees around the circle it would look exactly the same. One region is marked already, to show how it works.

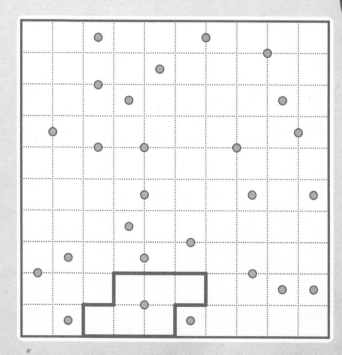

HERD IMMUNITY

Anti-vaccination misconceptions have led to worryingly widespread declines in the rates of vaccination for many diseases in many parts of the world. But why is this so dangerous for the public health of everyone? It is because one of the main ways that vaccines protect us is through what is known as herd immunity.

Disease agents – such as the germs from measles – are only contagious for limited periods. After a while, a person with the disease either dies or recovers and eliminates the infection. If there are enough potential hosts in a population, the germ can circulate indefinitely and become endemic, moving on to new hosts before it has worn out its welcome elsewhere. But if enough of the population – aka the herd – are immune to the disease, the maths does not add up for the germ, and it simply runs out of new victims to host it.

Therefore, if enough of the population has been vaccinated it creates herd immunity, meaning even those who have not been vaccinated are protected.

HOW HERD IMMUNITY WORKS

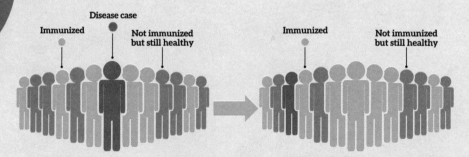

POPULATION NOT IMMUNIZED

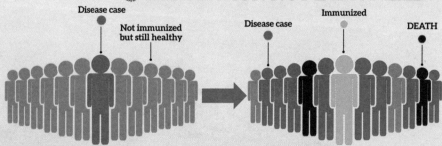

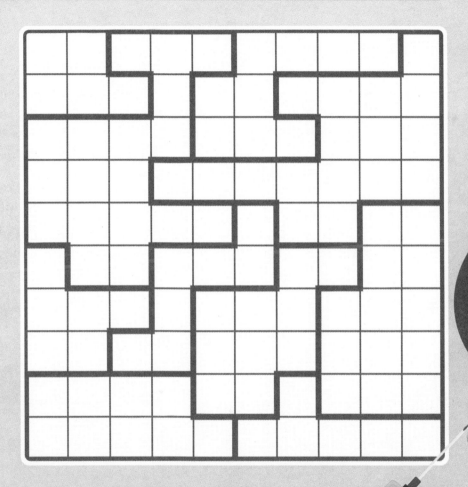

PUZZLE INSTRUCTIONS
Isolate the unvaccinated subjects so that there are just two in each area, and they don't spread disease by touching. Place the people in some squares so that each row, column and outlined region contains exactly two people. They cannot touch each other – not even diagonally..

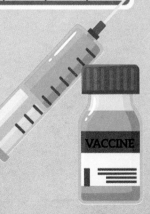

Even before the Covid-19 pandemic, humans have long had a gory fascination with the most disgusting and terrifying diseases. At the top of most people's lists would be horror diseases such as Ebola and other haemorrhagic fevers. These are highly contagious and have very high mortality rates, causing terrible and painful symptoms, such as bleeding from various orifices all over the body. Another gory candidate is *necrotizing fasciitis*, a bacterial infection more commonly known as the "flesh-eating disease". It spreads with shocking speed and causes horrific lesions.

But there is a case to be made that the world's worst diseases are actually those that are the most common and widespread, and thus quietly cause the greatest sum of human misery. Malaria, for instance, though rarely fatal is estimated to cost the African economy $12 billion every year because of productivity losses. Worse, easily treated and preventable diarrhoea kills 2,195 children every day – more than AIDS, malaria and measles combined!

HEALTH

PUZZLE INSTRUCTIONS

Track the spread of disease from each house in the village. Draw a single horizontal or vertical line across the full width or height of the centre of every white cell. The total length of all unbroken lines touching each blue square must be equal to the number printed on that square. The length of a line is equal to the number of squares it covers.

		3		5				
	7				1			1
			1				4	
3				6				
		1				7		
				5				3
	8				3			
1			2				3	
			5		6			

DEATH MAP

A crucial landmark in public health and medicine was the 1854 death map created by Dr John Snow. This was a map that showed the Soho area of London, on which were marked the residences of people who had died in an outbreak of cholera. At the time, outbreaks of cholera were ravaging the rapidly growing and industrializing cities of Europe, but little was known about the causes of the disease.

Snow's map showed that the cases of the recent cholera outbreak where clustered around, or could be traced back to, a single communal water pump. This revealed that a water-borne agent of infection was probably to blame. Identifying contamination of drinking water with human sewage as the likely cause of the disease eventually led to the introduction of proper sewage systems, deemed to be the most successful public-health intervention of all time.

▲ John Snow has received one of the highest honours that can be bestowed – a popular pub is named after him in Soho.

PUZZLE INSTRUCTIONS

The water pumps are missing from the map. Can you add them back in based on the death counts? Place the water pumps into some empty squares in the grid. Clues in some squares show the number of water pumps in touching squares – including diagonally. No more than one water pump may be placed per square.

3		3	1	2			
					3	4	2
3	3			3			
		1			5	3	2
3	3	2			3		
			3			3	2
3	3	3					
			1	2	1		1

HEAD TRANSPLANTS

Although met with almost universal censure and clinical incredulity, Italian neurosurgeon Sergio Canavero made headlines in 2016 when he claimed to be about to perform the first head transplant. Canavero claimed to have proven, via experiments with cadavers, that electrical stimulation and biological glue would achieve what every other expert insists is impossible: connecting severed nerve endings together to restore neural pathways.

In what was likely a fortunate turn of events for all involved, the controversial surgery never happened, but the plan did raise some intriguing philosophical questions. For example, would it not be more proper to speak of a body transplant, since the head is effectively the recipient not the donor? And what effect would a new and foreign body have on the psyche of the recipient? Might we, perhaps, discover that embodiment is a key aspect of consciousness, so that a different body would create a new and different consciousness for the recipient?

As the science of medicine progresses further, the likelihood is that what now seem to be madcap proposals – such as this one – will become commonplace. When this occurs, ethical questions such as the above will become more important to consider before medical actions are taken.

◀ Frankenstein's model in Madame Tussauds. They try to use fewer metal bolts in transplants nowadays.

HEALTH

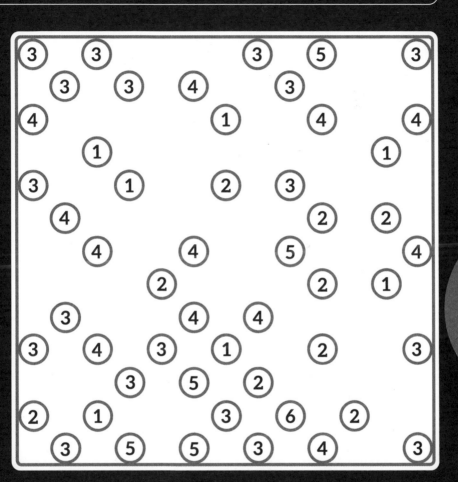

PUZZLE INSTRUCTIONS

To complete the head transplant successfully, you must make various neuronal connections across the divide from the severed brain stem to the spinal cord. To do so, join circled numbers with horizontal or vertical lines. Each number must have as many lines connected to it as specified by its value. No more than two lines may join any pair of numbers, and no lines may cross. The finished layout must connect all numbers so that you can travel between any pair of numbers by following one or more lines.

IS 10,000 STEPS ENOUGH?

Step counting has become one of the cornerstones of modern healthy lifestyle advice, and the magic target that is almost universally touted is 10,000 steps a day.

It is true that there have been several studies showing that people who walk 10,000 steps a day are healthier than those who walk fewer steps. However, the 10,000 figure has very arbitrary origins: it was plucked out of the air for marketing purposes by a Japanese company when it was bringing to market the first step counter. Importantly, it is not backed up by research.

Conversely, some studies have been made into the walking habits of modern hunter-gatherers, who live what health gurus call a "paleo" lifestyle and suffer few of the health issues that plague people in developed countries. For example, an investigation was made into the Hadza people in Tanzania, and found that they actually averaged 15,000 steps a day. This was backed up by a study of postal workers in Glasgow, Scotland, who similarly averaged 15,000 steps a day and enjoyed significantly better cardiovascular health than their more sedentary peers.

◀ All you need to do is take 10,000 steps a day and you will look like this too…

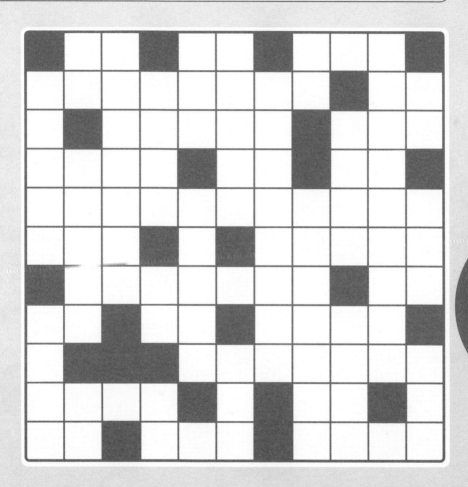

PUZZLE INSTRUCTIONS

Find the longest walking path possible, without visiting any location more than once. Draw a loop which visits as many white squares as possible, without visiting any square more than once. The blue squares cannot be travelled through, and the loop must be made up only of horizontal and vertical lines.

ANTIBIOTIC RESISTANCE

The Centers for Disease Control and Prevention (CDC) in the United States calls antibiotic resistance "one of the biggest public health challenges of our time" because it threatens to make obsolete the most potent therapies available to medicine. Antibiotic resistance occurs because natural variation means that some bacteria will be more resistant than others to antibiotics. Widespread use of antibiotics increases the number of these resistant strains, especially when antibiotics are used carelessly and at sub-lethal levels (as when added to livestock feed, for example). As a result, dozens of potential disease-causing germs are now difficult or impossible to treat with antibiotics.

The problems with this are obvious, and potentially devastating. In 2016, for instance, 490,000 people developed tuberculosis that was resistant to multiple drugs. HIV and malaria are also developing drug-resistant strains. Another common antibiotic-resistant germ, mostly encountered in hospitals, is MRSA (methicillin-resistant *Staphylococcus aureus*); people with this strain are 64 per cent more likely to die than people with a non-resistant form of the infection. In the US alone, nearly 3 million people a year acquire an antibiotic-resistant infection, and more than 35,000 people die from them.

▼ A 3D illustration of the MRSA bacteria.

V				A		A	
	V					V	
		A		A		V	V
	V			A			A
A			A			V	
V	V		V		A		
	A					A	
	A		A				V

PUZZLE INSTRUCTIONS

Balance the (V)iruses and (A)ntibiotics so that the the viruses don't take over more than two squares in a row at once. Do this by placing a V or an A in every empty cell so that there are an equal number of both letters in each row and column. Reading along a row or column, there may be no more than two of the same digit in succession. For example, AVVAAVVA would be a valid row but AVVVAAVA would not be valid due to the three Vs in succession.

THE HEALTH QUIZ

1) Who discovered the antibiotic properties of penicillin?

a) Alexander Fleming ☐

b) Alexander Armstrong ☐

c) Ian Fleming ☐

2) Which part of the eye perceives light and generates nerve signals to send via the optic nerve?

3) Which blood type is known as the "universal donor"?

4) From which organ does the aorta transport blood?

5) Is influenza caused by a bacterium or virus?

6) Which disease is believed to have caused the Black Death in the Middle Ages?

7) What colour of blood cell mediates the human immune response?

8) Which is the human body's biggest organ?

9) How many teeth does an adult human have?

10) What does the acronym ECG stand for?

11) Which fire-retardant fibrous material can cause mesothelioma cancer in the lungs?

12) How many chromosomes are there in a typical human cell nucleus?

13) What is the biggest bone in the human body?

14) Against which diseases does the MMR vaccination protect?

15) What are the letters that make up a DNA sequence?

16) Which toxic metal was used as a treatment for syphilis in the Middle Ages?

17) Roughly how many times does the human heart beat in an average lifetime?

a) 400 million ☐

b) 4 billion ☐

c) 400 billion ☐

18) Roughly how many bacteria live in your intestines?

a) 1 million ☐

b) 100 million ☐

c) 100 trillion ☐

19) Which organ of the body makes insulin?

20) Where would you find renal calculi?

HEALTH

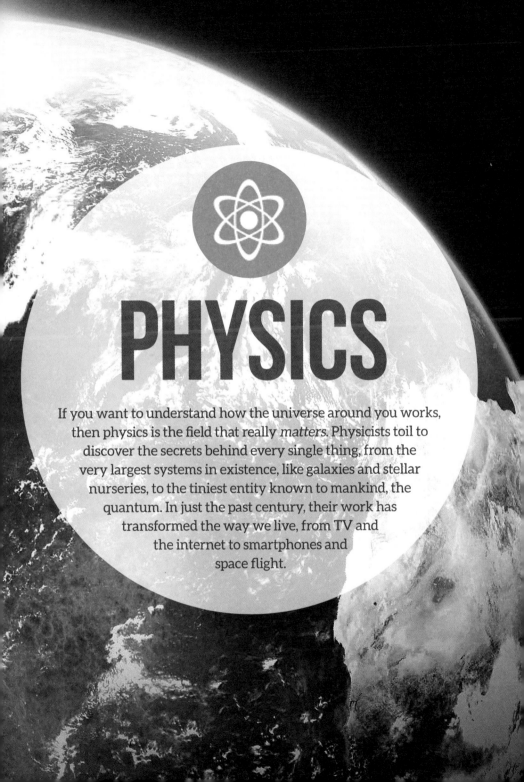

PHYSICS

If you want to understand how the universe around you works,
then physics is the field that really *matters*. Physicists toil to
discover the secrets behind every single thing, from the
very largest systems in existence, like galaxies and stellar
nurseries, to the tiniest entity known to mankind, the
quantum. In just the past century, their work has
transformed the way we live, from TV and
the internet to smartphones and
space flight.

Up until the sixteenth century, scholars believed the ancient Greek theory of falling objects, which held that heavier things fall faster. Although intuitively appealing, this implied a paradox that was pointed out by Galileo: what would happen if a heavier object were bound to a lighter one? On the one hand, the lighter object should fall more slowly and thus hold back the heavy one; but on the other hand, the combined object would be heavier than its component parts, and thus fall faster than either. The only solution to the paradox, Galileo explained, would be for both objects to fall at the same rate.

To prove his contention that objects can fall at the same rate independently of weight, Galileo was said to have dropped cannonballs of the same size – but different weights – from the Leaning Tower of Pisa, which showed that they landed at the same time.

✏ PUZZLE INSTRUCTIONS

Find the heights of these different buildings to test your theory from. Place 1 to 5 – representing the heights of buildings, where 1 is the shortest and 5 is the tallest – once each into every row and column of this "skyscraper" grid. The digits must be placed in the grid in such a way that each given clue number outside the grid represents the number of "buildings" that are "visible" from that point, looking along that clue's row or column. A digit is visible unless there is a higher digit preceding it, reading in order along that row or column. For example, in "21435" the 2, 4 and 5 are visible, but 1 is obscured by the 2, and 3 by the 4.

PHYSICS

▼ Galileo: a notorious cannonball-throwing medieval menace.

SCHRÖDINGER'S CAT

Discoveries in quantum physics in the first half of the 20th century, such as Heisenberg's Uncertainty Principle, implied that particles could be indeterminate. This is quite an extraordinary idea, because it means a particle might exist in one form or another simultaneously, only adopting a single form once observed (see *The Slit Experiment* overleaf). This concept is difficult to imagine, but when discussing an electron or a photon, these ideas have profound implications for our conception of reality. Accordingly, in 1935, Austrian physicist Erwin Schrödinger suggested a thought experiment to illustrate the remarkable implications of quantum mechanics to an everyday audience.

He devised a way of linking indeterminacy in a subatomic particle to indeterminacy at the scale of everyday life, dreaming up what he called "an infernal device". The device was a box containing a cat, a vial of poisonous gas, a hammer that might fall on the vial, and a trigger mechanism that could be set off through detection of the emission of a single particle of radiation from a grain of radioactive material. The decay of the radioactive matter is governed by the rules of quantum physics, so that the emission of the particle and consequent death of the cat are also governed by the laws of indeterminacy. Only by opening the box and observing the state of the cat is the indeterminacy resolved. Until this happens, the cat is literally both alive and dead at the same time!

PUZZLE INSTRUCTIONS

The Xs and Os represent cats – dead and alive! Observe whether each box has a dead cat or a living cat as you solve the puzzle. Place either an X or an O into each empty square so that no lines of four or more Xs or Os are formed in any direction, including diagonally.

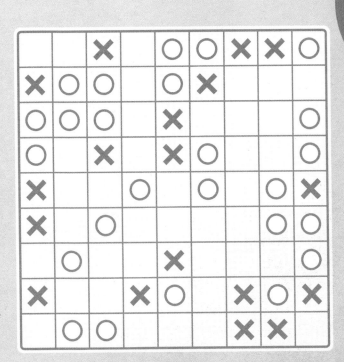

THE SLIT EXPERIMENT

A simple – but mind-bending – experiment shows that light acts simultaneously as both a particle and a wave, something that is intuitively impossible. If you fire photons (i.e. shine a light) at a slit in a barrier and record the pattern of impacts on a screen on the other side, you will see a single narrow bar of light, as expected from a particle.

However, if you open a second slit in the barrier, the pattern produced on the screen instead shows what you would expect from waves interfering with each other, rather than particles. Bright bars appear where the peaks of the waves coincide (as shown in the diagram), even though you are only firing one photon at a time at the slits.

This is strange enough, since it seems that the light has switched its form from particulate to wavelike, and the single photon is somehow passing through both slits at the same time! Even stranger is that, if you install a detector at the slits to detect through which slit the photon passes, the photons immediately stop behaving like waves and revert to behaving like particles. The simple act of observing photons actually changes their behaviour.

Photon: a particle representing a quantum of light or other electromagnetic radiation.

Wave: a kind of oscillation that travels through space and matter.

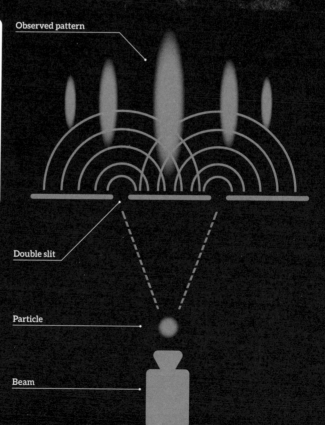

Observed pattern

Double slit

Particle

Beam

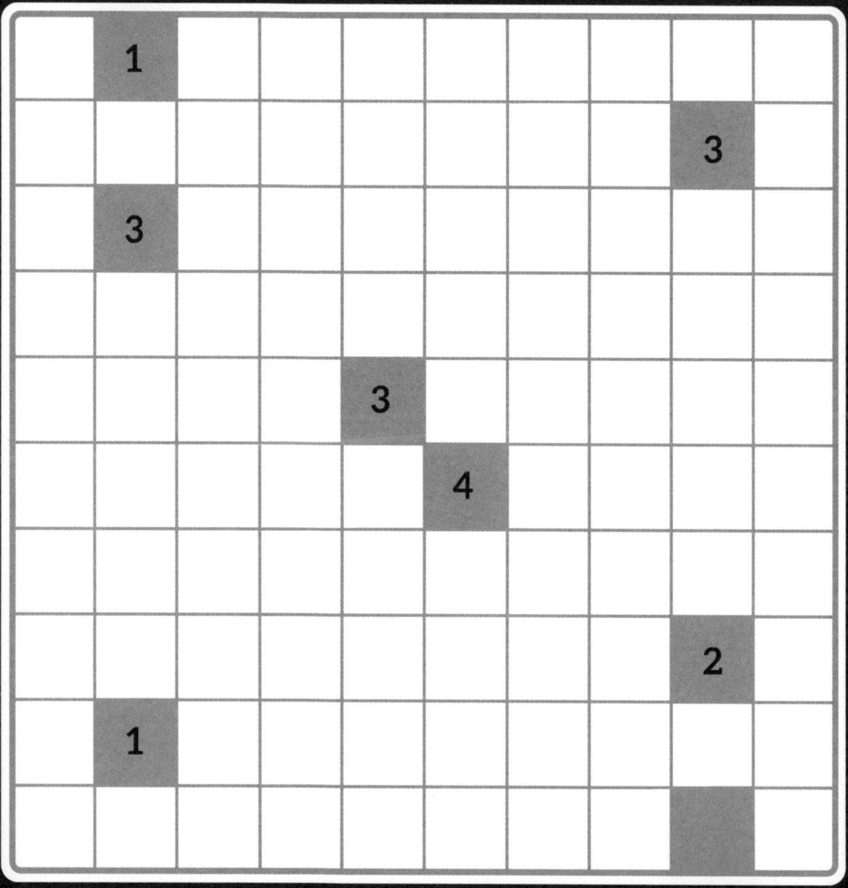

PUZZLE INSTRUCTIONS

Install the light emitters in the correct places to complete the experiment. Place light bulbs in some empty cells so that each empty cell is illuminated. Light bulbs illuminate every cell in their row and column until blocked by a blue cell. No light bulb can be illuminated by another light bulb. Some blue cells have numbers, which correspond to the number of light bulbs in their adjacent, non-diagonally touching, cells. Not all light bulbs are necessarily clued. All blue cells are given.

TRAVELLING AT THE SPEED OF LIGHT

Since his days as a student, Albert Einstein was puzzled by an apparent paradox relating to the speed of light. Common sense dictates that the velocities of things are relative: if you are on a train travelling west at 100 mph, but you are moving down the train in an easterly direction at 10 mph, relative to someone standing on a platform as you pass by, your net velocity is 90 mph toward the west. Similarly, if, while standing still and aiming west, you fired a gun, the bullet would not travel as fast as if you had been standing on a vehicle that was travelling west when you fired. These are both versions of what physicists call Galilean relativity.

What flummoxed Einstein was that James Clerk Maxwell's equations governing electromagnetic waves had proven that the speed of light in a vacuum (known as c) is a constant. A laser beam shot by a stationary person moves at the same speed as one shot by an astronaut on a spaceship travelling at 1,000 mph. How could this be possible?

Einstein's Theory of Relativity show that the constancy of the speed of light means that time and space are relative. One of the many results of this is that two people moving relative to one another at speeds close to c, for example, will experience time differently, and would each see the other's wristwatch moving more slowly than their own.

▼ Einstein was only briefly flummoxed by Galilean relativity. However, the use of a comb eluded him his whole life.

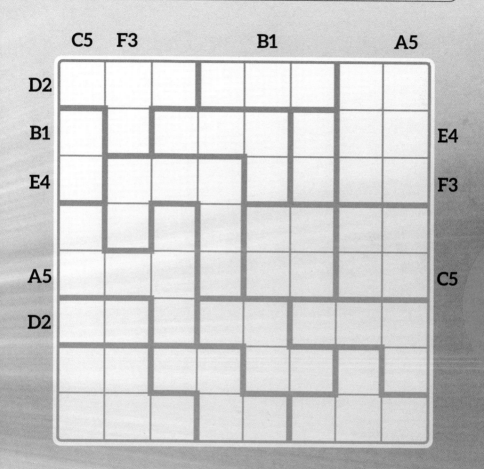

PUZZLE INSTRUCTIONS

Fire lasers at the speed of light into the grid. Draw diagonal lines across certain squares to form mirrors, with exactly one mirror per bold-lined region. The mirrors must be placed so that a laser fired into the grid from each lettered clue would then exit the grid at the same letter elsewhere, having bounced off the exact number of mirrors indicated by the number next to the letter.

SPLITTING THE ATOM

Nuclear fission is the process whereby a large, unstable atomic nucleus – for example the nucleus of an atom of radioactive uranium – splits into two smaller nuclei, releasing a large amount of energy in the process. The concept was originated by German chemist Lise Meitner and her nephew, the physicist Otto Frisch, while trying to explain why uranium atoms bombarded with neutrons appeared to have changed into much smaller atoms of barium.

The two scientists worked out that a single such event must release around 200 million electron volts (MeV) of energy. For the sake of comparison, the most energetic chemical reactions release approximately 5 eV per atom. They were even able to show where this massive burst of energy comes from: in the process of splitting from a uranium nucleus into smaller nuclei, a tiny amount of mass disappears. Using Einstein's formula $e=mc2$, Meitner and Frisch calculated that the missing mass equated to 200 MeV of energy.

NUCLEAR FISSION

Energy

Neutron

Neutron

Neutron

Fissionable Nucleus

Chain Reaction

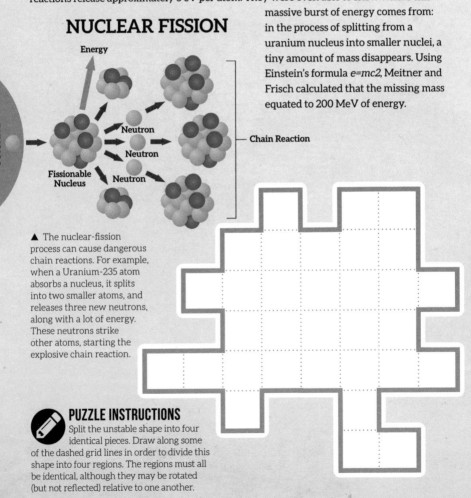

▲ The nuclear-fission process can cause dangerous chain reactions. For example, when a Uranium-235 atom absorbs a nucleus, it splits into two smaller atoms, and releases three new neutrons, along with a lot of energy. These neutrons strike other atoms, starting the explosive chain reaction.

PUZZLE INSTRUCTIONS

Split the unstable shape into four identical pieces. Draw along some of the dashed grid lines in order to divide this shape into four regions. The regions must all be identical, although they may be rotated (but not reflected) relative to one another.

STRING THEORY

Einstein's Theory of Relativity and some of the discoveries made in the field of quantum physics are, unfortunately, incompatible with each other. Cosmologists working to reconcile relativity and quantum physics are seeking a "Theory of Everything". One possible set of solutions involves imagining the existence of other dimensions in addition to the four-dimensional space-time of relativity – up to seven or more extra dimensions.

The obvious question is why these dimensions aren't visible to us. One suggestion is that they are folded up into incredibly tiny multi-dimensional string, and that what we perceive as sub-atomic particles are actually vibrating loops of this string. Current technology is not able to detect this potential string, making the theory very hard to test or falsify.

STRING THEORY

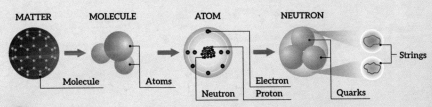

PUZZLE INSTRUCTIONS

Locate the set of vibrating strings in the grid. Draw a set of loops that together pass through all squares, using all of the given fragments. In each square a loop may pass straight through, turn ninety degrees or cross directly over another loop segment. Apart from in crossing squares, only one loop may enter any one square. Lines can never be drawn diagonally. Each loop must pass through at least one circle, and all circles with the same number must be part of the same loop and no other.

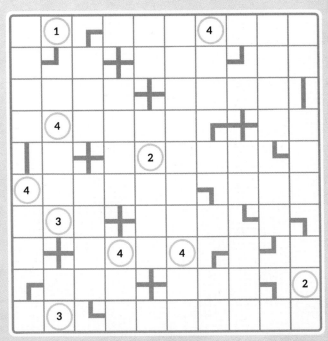

NEWTON'S APPLE

The legend of Newton's apple has become so familiar that it has now become hopelessly diffuse; a cliché in which the idea of gravity strikes the scientist on the head as he reclines beneath an apple tree.

In fact, the tale comes from a story told to relatives by Newton himself, in which he was musing upon the nature of gravity when he observed the fall of an apple from a tree in his garden. Gravity was, of course, a universally observable and known fact. What was not properly understood was why objects fell, and what laws governed this force. The falling apple prompted Newton to wonder if the same force that acted on the apple might extend out as far as the moon, thus explaining why it orbited the Earth instead of flying off into space.

He further speculated on whether he could work out the relative magnitudes of the force of attraction acting on both apple and moon. Using what he knew of the distances involved, Newton was able to figure out that they could, indeed, be one and the same force, and even managed to show that the strength of gravity varied according to the inverse square of the distance between the Earth and the apple or moon.

▼ Isaac Newton's laws of motion explained both the movement of objects on a small-scale, like apples falling from trees, and on a very large-scale, like planetary motion and the movement of oceans through tides.

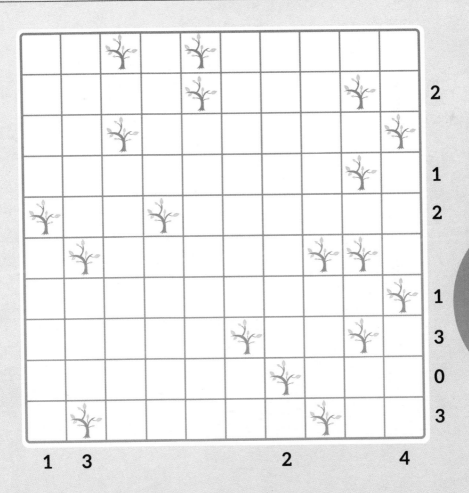

PUZZLE INSTRUCTIONS

Place the apple that has fallen from each tree. Attach exactly one apple to each tree, so that every row and column contains the given number of apples. Apples can only be placed in one of the (up to) four empty squares immediately above, below or to the side of a tree. No two apples can touch, however – not even diagonally.

RAINBOWS AND BLUE SKIES

The colours of rainbows – and, in fact, the sky itself – are the result of components of the atmosphere absorbing or scattering light as it passes through them. Light coming from the sun is a mixture of all wavelengths (although, apparently, when viewed from space, the sun is a peach-pink colour). To reach your eyes, however, light must pass through the atmosphere. Air molecules scatter short-wavelength blue light, but transmit red and yellow light. This means that the sun itself looks yellow, while the blue light reaches your eyes by a much more roundabout route, so that every other part of the sky looks blue. A rainbow is seen when light is refracted by water droplets in the air, which separate the wavelengths in the same way that a prism splits white light into a spectrum.

PHYSICS

START

FINISH

 **PUZZLE INSTRUCTIONS**
Find your way across the rainbow-shaped maze. Can you travel across the rainbow, from the left-hand to the right-hand edge?

THE COLDEST PLACE IN THE UNIVERSE

Instinctively, you might expect that the coldest place in the universe would simply be "outer space". Indeed, the average temperature of space is a chilly 3 K (-454 °F, -270 °C). "K" stands for "kelvin", which is the unit favoured by scientists for measuring temperature, because it is absolute and it starts at zero; 0 K is known as "absolute zero". This is the state in which no energy at all is present, and it is also purely theoretical since it cannot exist in reality. However, the closer you get to it, the stranger physics becomes, and so scientists are very interested in approaching as near as possible to absolute zero purely to analyse what happens.

Some parts of space are particularly cold. The coldest known part of space is the Boomerang Nebula, which has a temperature of just 1 K! However, using lasers and magnets, scientists have been able to hold atoms so still that their temperature drops below even this. The current record for the coldest temperature ever achieved (and, therefore, the coldest place in the universe) is inside the Cold Atom Lab, a device on the International Space Station, which has achieved a record temperature of 100 nanokelvin – one ten-millionth of a kelvin above absolute zero.

◀ The Boomerang Nebula, as captured by the Atacama Large Millimeter Array in June 2017.

PUZZLE INSTRUCTIONS

Plot a scientific path to 1 ten-millionth of a Kelvin, starting at 64 ten-millionths and counting down. Write a number in every empty square, so each number from 64 to 1 appears once in the grid. Numbers must be placed so as to form a path from 64 to 1, moving to a square one lower in value at each step exactly as a king moves in chess – i.e. left, right, up, down or diagonally.

61	58	59			46		
				49			
	64				48		
	52					36	
			1			33	
	26		17				
	24		15		4	7	8

THE BIGGEST SCIENTIFIC EXPERIMENT IN HISTORY

The Large Hadron Collider (LHC) at CERN (the European Organization for Nuclear Research) offers a list of superlative records: it is the biggest scientific experiment, the biggest detector and the biggest machine in history. It is, quite simply, a colossal particle accelerator (hadrons include protons and neutrons), running around a giant circular tunnel. It lies 100 ft (33 m) down and 27 km (16.5 miles) long, straddling the border between Switzerland and France, near Geneva.

Nearly 10,000 magnets accelerate charged particles around this loop at 99.99 per cent of the speed of light until they smash into each other with colossal force, recreating the conditions one micro-second after the "Big Bang". The magnets of the LHC weigh more than the Eiffel Tower, while the entire device weighs more than 38,000 tonnes – over half the weight of the *Titanic*. Because the magnets need to be super-cooled to be super-conductive, the LHC also features the world's largest refrigerator – just to sneak another record in there.

▼ A small part of the very Large Hadron Collider, in France.

3			5				
					8		
	6						
		4			7		
4						5	
		4					
							3
2				5			

PUZZLE INSTRUCTIONS

Create a loop which passes by each numbered "magnet" as detailed by its configurations. Draw a single loop that passes through some of the empty squares, using horizontal and vertical lines. The loop cannot re-enter any square. The loop must pass through the given number of touching squares next to each numbered "magnet", including diagonally touching squares.

111

SPOOKY ACTION AT A DISTANCE

One mystery of quantum physics that disturbed Einstein – who called it "spooky action at a distance" – is quantum entanglement. This is a property of pairs of particles, which follows on from Heisenberg's indeterminacy principle. When the pair of particles is created, some of their properties will be opposite, so that, for example, if one of the pair has spin in one direction, the other must have spin in the other direction.

So far, so simple. Where it becomes confusing is that which particle has which spin is indeterminate, until at least one of the particles is observed. This determines its version of the property, and thus, by extension, the property of the opposite particle. The two particles are said to be entangled. What is spooky is that, if the particles are separated before being observed, and then one of the particles is observed, the other one, despite being physically distant, immediately becomes determinate. It is as though the first particle has "told" the second one in which direction to spin (for instance), and this information appears to travel instantaneously – faster than the speed of light.

A number-path puzzle grid containing the following values (read left to right, top to bottom):

Row 1: 10 — 10 — 10 — 4 — 4 — 11 — 13
Row 2: 10 10 — 3 — 3 — 2 7 — 7 — 2 — 4 — 4
Row 3: 8 — 4 6 — 6 — 1 2 — 6 — 6 — 2 5 — 2 — 11
Row 4: 8 — 2 — 4 — 5 — 5 — 4 — 4 — 2 4 1 4
Row 5: 10 — 2 2 — 1 4 — 4 — 2 — 2 2
Row 6: 6 — 2 — 2 — 4
Row 7: 6 — 5 — 4 4
Row 8: 4 — 6 — 2
Row 9: 4 — 5 — 5 — 1 6 7 — 7 — 2 4 — 4 4
Row 10: 9 — 10 — 11 — 11 4 — 4 — 1 13
Row 11: 9 — 5 8 — 8 5 10 — 2 2 — 8 — 8 — 1 — 1 4
Row 12: 5 — 5 — 3 — 3 — 4
Row 13: 2 2 — 3 — 3 2 — 2 — 4
Row 14: 5 — 4 — 4 — 9 — 2 4 — 2 7 — 4 — 9
Row 15: 2 4 — 4 5 — 1 — 2 4 — 4 — 4 — 3 1 — 4 — 2 2 — 11
Row 16: 2 — 2 2 — 4 — 2 — 2 2 — 4 — 3 — 7 — 5
Row 17: 9 — 3 — 4 — 4 — 4 — 3
Row 18: 9 — 4 — 4 3 — 4 — 5 — 5 — 7 — 5 3
Row 19: 12 — 5 — 6 — 13
Row 20: 7 — 4 — 9 2 — 6 7 — 11
Row 21: 2 2 — 2 — 6 — 9
Row 22: 5 — 4 4 — 4 — 4 — 4 — 10 4 — 7 2
Row 23: 4 4 — 4 — 2 2 — 4 4 — 5 — 2 — 12
Row 24: 6 6 — 3 — 4 — 5 — 5
Row 25: 12 — 5 — 5 — 3 — 14 — 3 — 3 — 4 — 4 6
Row 26: 7 — 2 — 7 — 14 6 — 7 — 6 — 8
Row 27: 1 — 4 — 2 — 5 — 5 — 6 — 4 5 — 1
Row 28: 4 — 1 — 5 — 4 4 — 2 2 — 7 7 4
Row 29: 5 — 5 5 2 2 — 4 — 7 — 7 — 2 — 7
Row 30: 4 — 2 — 2 2 — 6 4 — 10 — 12 — 4
Row 31: 4 — 1 2 — 2 2 — 4 5 — 2 2 — 5 — 9 — 12 4 4 6 — 4
Row 32: 4 — 2 — 2 — 4 — 4 3 2 — 3 — 8
Row 33: 5 6 — 3 — 2 — 3 11 — 2 2
Row 34: 6 — 4 4 — 1 — 3 — 3 — 9 — 5 — 13 12
Row 35: 4 — 9 — 9 — 5 — 1 10 — 6
Row 36: 9 5 — 4 — 4 — 4 — 4 — 5 — 6
Row 37: 9 — 6 5 — 3 — 3 — 4 — 4 — 5 — 5 6
Row 38: 6 — 5 — 5 6 — 6 4 — 4 — 2 2 11 — 10 13

PUZZLE INSTRUCTIONS

Join the pairs of particles to reveal a picture. Draw paths to join numbers into pairs of identical numbers, where each path visits the number of squares equal to the numbers it joins, including those at either end. Paths can only move horizontally or vertically between squares, and no more than one path can enter any square. Once complete, shade all squares containing paths to reveal a hidden image.

FRIDGE MAGNETS VS THE EARTH

Thanks to its metallic core, the Earth has a vast magnetic field that extends around 70,000 km (43,500 miles) out into space, although if conditions are right, this can balloon to over 300,000 km (186,400 miles). But the strength of the field is surprisingly puny. A fridge magnet is roughly 20 times more magnetic than the Earth. Magnetic-field strength ("magnetic pull") is measured in tesla (T) and gauss (G): 1 T = 10,000 G. The Earth's field strength is roughly 0.5 G, although it varies across the planet – it is strongest at the poles and weakest at the equator. A fridge magnet is around 10 G, while the electromagnet used in an MRI machine could be up to 3 T (30,000 G). This is a full 3,000 times more powerful than a fridge magnet and 60,000 times more powerful than the Earth's magnetic field.

PUZZLE INSTRUCTIONS

Work out where the magnets are based on the readings given. Each oval must either be shaded to show there is no magnet there, or marked with + at one end and – at the other to indicate its polarities. Same magnet polarities cannot touch to the left/right/above/below. Numbers outside the grid reveal the total number of +s and –s in each row and column.

PHYSICS

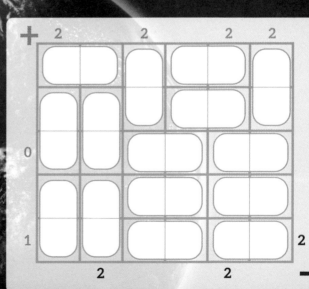

TIME-TRAVELLING TWINS

Einstein's theory of relativity has many mind-bending ramifications. One (as we've already seen) is that travelling at great speed, relative to someone standing still, causes time to pass more slowly for the traveller, as perceived by the sedentary observer. The thought experiment used to illustrate this involves a pair of twins. One is an astronaut that travels far out into space before returning to Earth, travelling at near-light speed. The other twin stays on Earth. When the first twin returns after what seemed to her like just a single year of travel, she finds that 20 years have passed for the Earth-bound twin. The Earth twin, meanwhile, was using a special telescope to observe the watch of the astronaut twin as she travelled in space and, to him, it appeared as though her watch was moving at one twentieth the speed of his own watch.

PUZZLE INSTRUCTIONS

Reunite the twins after so much time apart (or not that much time apart, depending on who you ask). Draw horizontal or vertical lines to join the circles into pairs, so each pair contains one shaded and one unshaded circle. Lines cannot cross either another line or a circle.

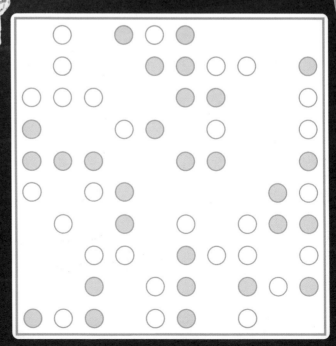

THE WEIGHT OF AIR

In 1641, Galileo took on a challenge posed to him by some Tuscan miners, who wanted to know why their water pumps would only draw up to a height of about 10 m (33 ft). If the tube of the pump were any taller than this, the column of water would separate from the pump's plunger, leaving an empty space at the head of the column. Received wisdom held that a vacuum was abhorrent to nature, yet it seemed that this must be one.

Galileo passed away soon after, but his secretary, Evangelista Torricelli, continued the investigation, creating the world's first barometer: a device for measuring atmospheric pressure, and thus going some way to predicting the weather. To make his experiment more manageable, Torricelli substituted dense liquid mercury for water. By filling a tube roughly a metre long with mercury, putting his finger over the end and inverting it in a dish of mercury, he could easily demonstrate a gap at the top of the tube, which must be full of nothing: a vacuum. The column of mercury that rose around 76 cm up the tube was held up, he explained, by the weight of the column of air in the atmosphere overhead. As the temperature and density of the column of air changed with the weather, so the precise level of mercury in the barometer would vary.

▼◢ A statue of Evangelista Torricelli next to a 19th century illustration showing how his early barometer worked to measure atmospheric pressure.

PHYSICS

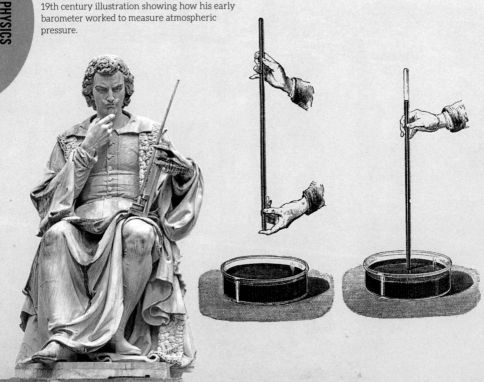

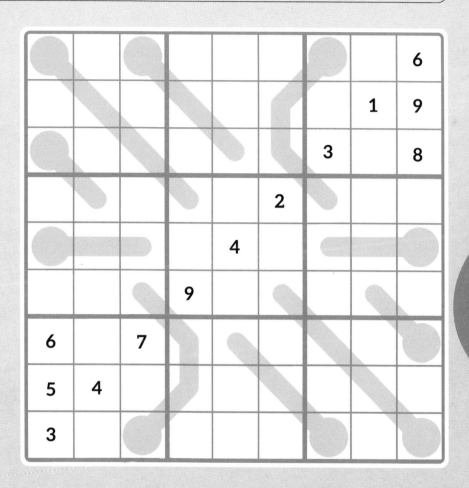

PUZZLE INSTRUCTIONS

Complete this barometer Sudoku. Place
1-9 once each into every row, column and
bold-lined 3×3 box. The value of the digits along each
shaded barometer must increase cell by cell from the
bulb (lowest value) to the head (highest value). This also
means that digits cannot be repeated in a barometer.

THE PHYSICS QUIZ

1) Which will fall faster in a vacuum: a feather or a cannonball?

2) What type of particle carries electromagnetic force?

3) Three containers are filled with boiling water. One is made of copper, one is made of cork and the third is made of ceramic. Which container will be the hottest to hold?

4) Which part of the spectrum has a slightly longer wavelength than red light?

5) What type of energy is given to a ball by raising it to the top of a hill?

a) kinetic energy ☐

b) thermal energy ☐

c) potential energy ☐

6) A high-powered rifle fixed 2 m off the ground fires a bullet horizontal to the ground. At the exact same instant that the bullet leaves the muzzle of the gun travelling at 160 m/s, the shell-casing drops from the rifle. Which will hit the ground first?

7) In the nucleus of an atom, which sub-atomic particle carries a positive charge?

8) For the detection of which phenomenon did the LIGO detector team win the Nobel Prize in Physics for 2017?

9) What property is possessed by silicon and germanium, which makes them useful in electronics?

10) If you painted a wall with paint that absorbs light in the blue and red regions of the visible spectrum, what colour would it be?

11) Which British physicist proved the existence of the electron in 1897?

12) What name is given to a machine that travels by burning fuel, which it carries to generate exhaust that is expelled in one direction to produce a force in the opposite direction?

13) Force equals mass multiplied by which quantity?

14) If you fill a bath to the brim and drop into it a cube of gold measuring 5 cm on each edge, how much water will slop over the edge of the bath?

15) In Einstein's famous equation $e=mc^2$, what does "c" stand for?

16) What does LHC stand for?

a) Lepton Heavy Compactor ☐

b) Laser Heating Collimator ☐

c) Large Hadron Collider ☐

17) Spaceship A is twice as far away from the Earth as Spaceship B. How many times less will be the force of gravity it experiences?

18) Which phenomenon causes the change in pitch of a siren that you hear when an ambulance speeds past?

19) What is alpha radiation?

a) helium nucleus particles ☐

b) energetic photons ☐

c) neutrons ☐

20) The giant Super-Kamiokande detector in Japan is looking for which particles?

PLANTS AND ANIMALS

Our planet is unique – at least as far as we know so far – in that it is the only planet in the unfathomably vast universe on which we have found intelligent life. And what an amazing variety of life – even without the Loch Ness Monsters and yetis that the cryptozoologists would tell you are real. From extinct giant sloths as big as a bear and crows that use cars to crush nuts, to hallucinogenic flowers and water bears that can survive in space, the world is far more wonderful than we often give it credit.

CRYPTOZOOLOGY

Cryptozoology is the study of strange and unknown animals; the term derives from the Greek word for "hidden". Naming it as such also represents an attempt to turn monster-hunting into a serious scientific pursuit. It has its roots in the work of Belgian zoologist Bernard Heuvelmans, whose 1955 book *On the Track of Unknown Creatures* is considered to be the founding text of the field. Some of the book deals with creatures that mainstream zoology had long dismissed as folktales or legends, but had eventually been forced to recognize – animals like the gorilla, the Komodo dragon and the okapi. It was Heuvelmans' hope that proper scientific study might bring similar acceptance to such creatures as the Loch Ness Monster and the yeti. Unfortunately for him, in the years since, cryptozoology has become more of a pseudoscience than a real one, mired in hoaxes and conspiracy theories.

◀ Although the okapi resembles half a zebra, it is actually more closely related to the giraffe.

▼ Komodo dragons were first recorded by Western scientists in 1910.

PUZZLE INSTRUCTIONS

Dig up all of the folklore creatures. Find the entries below the grid, ignoring any spaces. They may be written in any direction, including diagonally, and may read either forwards or backwards.

```
F K E K A N A N S I E O M T A
S Q N O G A R D A N G S O S A
C A R E M I H C P P A Y K O H
S R R F W U Y M C S E P O N A
H B T B A K A R Q T T I L O R
N M N R A C E U I H T N E K A
I L D A O C A L U O G F M L W
H Y U P N T A N P R U I B A A
H H P M C D D P I I T B E I K
G I E H E E I F U S E I M B O
H E A I R R F B U H E G B H N
S U N B W I M M E H C F E P I
C G I I N O A A A A P O K A M
N R B O E D Y S I C R O E W T
D O E A E D I U R D N T M N O
```

ALKONOST
ANANSI
BIGFOOT
CHIMERA
CHUPACABRAS
DRAGON
GENIE

GRIFFIN
HIPPOCAMP
HYDRA
ITSUMADE
KELPIE
MERMAID
MINOKAWA

MOKOLE-MBEMBE
NANDI BEAR
SASQUATCH
THUNDERBIRD
YETI
YOWIE

HOW MANY SPECIES ARE THERE ON EARTH?

PLANTS AND ANIMALS

▲ Carl Linnaeus, the father of modern taxonomy. His body was also the first to be described as *Homo sapiens*, and so he is the type specimen for the entire human race!

▼ The major ranks in which every single species on Earth is classified.

Eukaryotes are organisms with relatively sophisticated cell structures, as opposed to prokaryotes, the group that includes single-celled organisms like bacteria. Since the eighteenth century, when the Swedish taxonomist Carl Linnaeus started to classify species, scientists have categorized about 1.6 million species of eukaryote. This total is believed to be just a fraction of the true number of species on Earth, which a 2011 study put at around 8.7 million.

It is expected that, in line with the number of known species, most of these will be animals, and that most of these animals will be insects. Of the known species, for example, well over half are insects. Most of the species we know about live on land, and it is estimated that about 91 per cent of marine species are unknown. Scientists fear that vast numbers of these unknown species will go extinct before we ever learn about their existence, a fear exacerbated by the increasing pace of climate change.

HIERARCHY OF BIOLOGICAL CLASSIFICATION

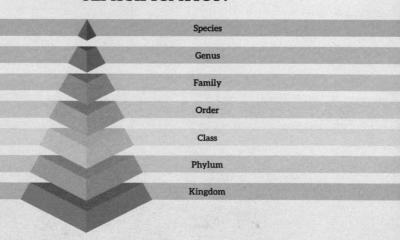

Species

Genus

Family

Order

Class

Phylum

Kingdom

PUZZLE INSTRUCTIONS

Reveal the unknown creature. Shade some squares according to the clues at the start of each row or column. The clues provide, in reading order, the length of every run of consecutive shaded squares in each row and column. There must be a gap of at least one empty square between each run of shaded squares in the same row or column.

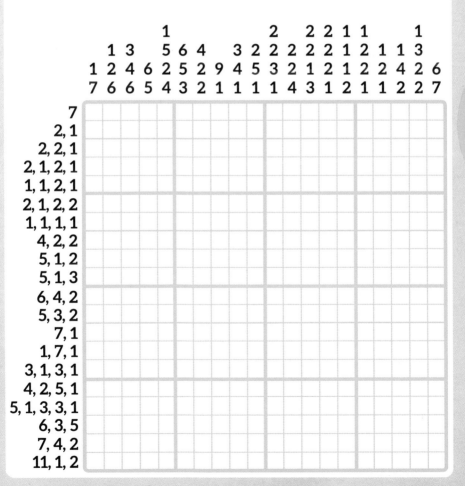

DARWIN'S MONSTER

Some organisms live fast and die young, but produce masses of offspring, which they generally leave to make their own way in the world. Biologists call these species r-selected, where the "r" is for "reproduction". Mice are a common example to think about. By contrast, k-selected species generally live much longer, tend to be much bigger, and invest a lot of effort and resources in raising their offspring, although they have very few of them. Humans fit this description, of course. Thus, r- and k-selection represent different life strategies for being evolutionarily successful.

In 1979, the biologist Richard Law imagined what he called "a Darwinian demon, which can maximize all aspects of fitness simultaneously"; in other words, an organism that is large, long-lived and heavily invested in its offspring, but also manages to breed very quickly and produce large numbers of offspring. Such a creature would quickly dominate its ecosystem, but fortunately it cannot exist in the real world, where resources are limited and where evolutionary fitness is about trade-offs.

▲ Charles Darwin was seen as something of a monster himself by the many who could not believed we shared common ancestry with apes.

PUZZLE INSTRUCTIONS

Place 1 or 2 "children" (mines) as needed next to each "person" (clue number). Place the children into some empty squares in the grid. Clues in certain squares show the number of children in touching squares – including diagonally. No more than two children may be placed per square.

2	3	3	2		1
			6		4
4	7		5		
		4		5	5
5	5			4	
		2	2		4

EDGE CASE

Slime moulds is the name given to a bizarre set of organisms that do not seem to fit into normal biological categories. They can exist as single-celled, amoeba-like organisms, but they can also gather together to form fungus-like multi-cellular structures. When two of the amoeba-stage cells meet, they sometimes merge to form a single cell, the nucleus of which then multiplies, while the cell itself simply grows bigger and bigger without dividing. Therefore, this organism pushes excitingly at the boundaries of what is normally understood to be meant by concepts such as cells, individuals and animals.

PUZZLE INSTRUCTIONS

Work out how much each organism has spread within the grid. Place a number in every empty square so that each number in the grid is part of a continuous region of that many squares. Regions are continuous wherever two squares of the same value touch. Two different regions made up of the same number of squares cannot touch. "Touch" in both cases does not include diagonal touching. You may find it helpful to draw solid borders around regions as you solve, to more easily see how they touch.

		2		3	2	7			
	5			7	3	7			
7	7			2	2	3		7	
	7					1		8	
1			8	2					
5				3		8	5		
7		7	5		3			2	
	1			7			1		
		1	4		1		3		
7		9							9

POISONOUS PLANTS

Plants make poisons for self-defence. Fortunately, if used correctly, many of these physiologically active substances can have pharmacological applications, from the pain-killing properties of willow and poppy extracts to the heart medicine that can be extracted from foxgloves. One of the main problems with plant toxicity, however, is that it varies greatly within species, and often even within the same plant. Toxins may be concentrated to different extents in different parts of the plant – in the castor-oil plant, for example, ricin is almost entirely concentrated in the castor bean, while in the *strychnos nux-vomica* plant, strychnine is present in the stem, bark and seeds. The knowledge of where the toxins are present is invaluable – not to mention potentially life-saving.

Toxicity varies over the lifetime of a plant – young pokeweed shoots, for instance, can be eaten in a salad, but berries and leaves from a mature plant contain poisonous lectins that can cause gastrointestinal problems. The fruit of the mandrake plant can be safely eaten when it is very ripe, but is extremely dangerous at other times. Mandrake is one of a family of toxic plants that includes belladonna, datura and henbane, which are all dangerously rich in what are known as belladonna alkaloids, such as atropine, hyoscine, hyoscyamine and scopolamine.

These are so hazardous that medical students are taught to remember the symptoms of belladonna alkaloid poisoning with a mnemonic: Mad as a hatter; Blind as a bat; Dry as a bone; Hot as a hare; Red as a beet.

◀ The foxglove is just one example of a common plant that can be highly toxic; in humans it causes nausea, vomiting and diarrhea. It has been used in centuries past by herbalists to help with weight loss, but this is very much *not* recommended!

PUZZLE INSTRUCTIONS
Find how much poison is in each area of the "greenhouse". Place a digit from 1 to 9 into each square, so that no digit repeats in any row, column, bold-lined 3×3 box or dashed-line cage. The numbers in each dashed-line cage must add up to the value given at its top-left.

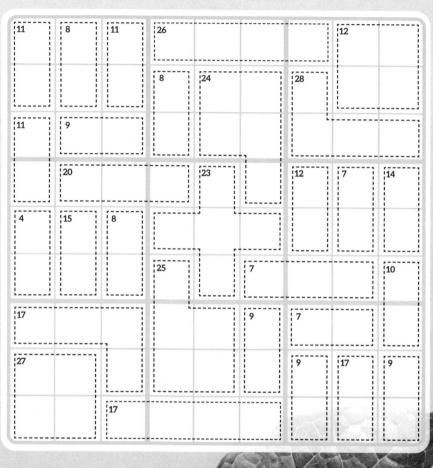

▶ The mandrake's roots and leaves are highly toxic, and even have hallucinogenic effects.

THE INDESTRUCTIBLE WATER BEAR

Water bears – also known as tardigrades – are microscopic eight-legged aquatic animals that can be found almost everywhere in the world, partly thanks to being so incredibly tough. They are among the few organisms to survive in Antarctica's McMurdo Valleys, thought to be the driest and coldest desert on Earth. In fact, tardigrades can survive incredible extremes, including being boiled in water, blasted with radiation, and chilled to the temperature of outer space.

Amazingly, they can even survive being *in* outer space, as a 2008 European Space Agency experiment demonstrated, when tardigrades were exposed to the harsh vacuum of space and to the radiation from which we are protected by the Earth's magnetic field. They achieve this feat thanks to their ability to go cryptobiotic – a state of super-hibernation, in which the tardigrade curls up into a tiny ball, sheds more than 95 per cent of the water in its body and replaces it with a kind of anti-freeze, before shutting down 99.99 per cent of its metabolic activity.

▼ The hardy – and highly attractive – tardigrade.

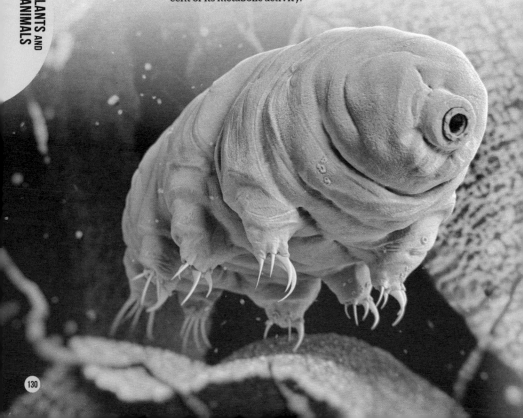

1							4		5		2		
2			5		4							2	3
	5	4	2		4	4	4	2		6			
	3		1	3	4	3		2	3				6
					7			4				7	
5				3			7	4	2		5		
	4	2	3					6	3	3			4
	4		3		6			9	6	3			4
	5	4	3		4	6		8					4
		3		1	2		4		5	3		1	
5	7				2		3	2					
4		4		2	4		4				1	4	
4		5			8				3		4		
	7						6		4			4	3
1		5	4		2		5		4		3	2	

PUZZLE INSTRUCTIONS

Reveal the creature from the numbers on the picture. Shade some squares, including some of those containing numbers. Each number reveals how many squares touching that number are shaded, including the number's square itself. When complete a simple picture will be revealed.

IS THERE LIFE IN EARTH?

New places are constantly being discovered which possess suitable conditions for life. A 10-year project called the Deep Carbon Observatory concluded in 2018 that the deep Earth harbours a vast store of almost entirely unknown life forms, nearly all of them bacteria and bacteria-like organisms known as archaea. It is estimated that around 70 per cent of all bacteria and archaea live underground, and organisms have been found at depths of up to 5 km below the surface. These creatures can survive an amazing range of extremes, from temperatures higher than the boiling point of water to crushing pressures hundreds of times greater than we experience on the Earth's surface. It is even believed that some of these cells could be tens of millions of years old.

What do they eat? Many are found in ocean-floor sediment, and probably eat organic fragments. However, deep-crust organisms might live on hydrocarbons such as oil, or have similar metabolisms to the bacteria that live in undersea vents, feeding on sulphur-rich compounds.

PUZZLE INSTRUCTIONS

Find the shapes of the various sediment areas. Shade some empty squares so that each number in the grid becomes part of an unshaded area of precisely that many squares. There must be exactly one number in each unshaded area. Shaded areas cannot cover a 2×2 or larger square, but all shaded squares must form one continuous area. Areas are not considered continuous if they only touch diagonally.

							3
					2		
			5				
				13		5	
	1		4				
					3		
			4				
3							

ECHOLOCATION

Echolocation – also known as bio sonar – is the use of sound for spatial perception and navigation. Bats are famously good at echolocation, emitting ultrasonic clicks and listening for the echoes to judge the position, range and motion of targets. They have evolved the facility to find insects while in flight at night, and they rely on it to help them catch the high numbers of bugs they need to feed their extremely active metabolism they use to power flying. A common pipistrelle – a tiny bat with a body just 3.5–5.2 cm (1.4–2.0 inches) long – can eat over 3,000 tiny insects in a single night.

Bats, however, are not the only species that can echolocate; others include whales, dolphins, shrews, some birds and even a few humans – some blind people have learned to navigate using echolocation.

◄ They might be blind as a bat, but it makes no difference to their hunting skills.

PUZZLE INSTRUCTIONS

Locate the bats based on the readings from various surveillance towers. Draw circles in certain squares to reveal hidden bats. Each green square is a tower, containing a number equal to the total number of bats in its row and column. Bats are one square in size, and cannot touch either another bat or a tower – not even diagonally. All bats can be seen by at least one tower.

	2						4
		1					
			0				
2		0					
				1			2
		2					
			2				
4					3		

WHO KILLED THE MAMMOTH?

The woolly mammoth and other large animals, ranging from the woolly rhino and giant sloth to the giant moa and the sabre-tooth tiger, are collectively known as Pleistocene megafauna. These large animals all began to die out around 60,000 years ago. The timing of the extinction of these animals correlates to the spread of anatomically modern humans around the globe. The obvious conclusion is that humans drove these creatures to extinction and that, in the popular conception, we over-hunted them.

We cannot know for sure if this is accurate. It's quite likely that it was not due to direct predation but instead competition for space and resources – such as water – and to human-caused changes to the ecosystem – for instance, through introduced species or landscape-burning. But it is not at all clear that humans were entirely responsible. There are other potential culprits – mainly climate change – and, in addition, there is evidence that humans and some megafauna co-existed for millennia, suggesting that some large animals survived the arrival of humans for some time.

▼ In true *Jurassic Park* style, scientists have inserted genes from the woolly mammoth into the genome of an elephant. There are fledgling plans to re-introduce them into Siberia!

PUZZLE INSTRUCTIONS

Can you discover all of the extinct animals? Find the entries that are below the grid, ignoring any punctuation or spaces. They may be written in any direction, including diagonally, and may read either forwards or backwards.

```
C J A V A N T I G E R G L H M
O N H T O L S T N A I G E B N
N O D O T O R P I D C U A P R
G B L U E B U C K G K M R E O
M I R Y O O N A L O W B D S G
A S A D B A R Y G E L G U M D
S M N N P L E R G A T R I I
T U A R T T I H G Z A A T L R
O S A M O D S B E I E U I O E
D T H D M E E L R B T E Q D W
O R O C L O L E E E O I S O O
N N T U O E T V R G S B L N L
T R T E D R A H B H A S G A F
T T I H E C U K N I M A E S B
D R D E N I C A L Y H T E L O
```

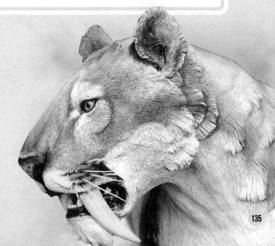

AUROCHS
BALI TIGER
BLUEBUCK
CAVE BEAR
DIPROTODON
DIRE WOLF
GIANT DEER
GIANT SLOTH
GLYPTODONT
JAVAN TIGER
LESSER BILBY
MAMMOTH

MASTODON
QUAGGA
RED GAZELLE
SEA MINK
SMILODON
TARPAN
THYLACINE
TULE SHREW

CLEVER CROWS AND PERSPICACIOUS PARROTS

Birds have small brains but evidence from both the lab and the wild shows they can demonstrate remarkable intelligence. Examples of avian smart thinking include carrion crows in Japan that place nuts on roadways, waiting for cars to crush the shells; green herons that pick up bread for ducks to use as fish bait; and blue tits that have learned which colour of milk-bottle top to raid for the creamiest milk.

In the lab, it is parrots and corvids (the bird family that includes crows) that have proven to be the smartest. Tests show, for example, that crows and rooks are better than eight-year-old children at reaching a treat by making a wire hook, while an African grey parrot called Alex was taught by Harvard psychologist Irene Pepperberg to understand and use approximately 100 words, and was able to understand concepts such as "same", "different" and "zero".

PLANTS AND ANIMALS

PUZZLE

On a particular housing estate, different birds have started to drink different types of milk from bottles left outside houses. The three birds that have been observed doing this are a blue tit, a robin and a blackbird. Each bird drinks a different type of milk with a different-coloured lid. The types of milk are 1% fat, 2% fat and 4% fat, and the lid colours are red, blue and silver.

1) The blue tit doesn't drink the fattiest milk, nor from the bottle with the silver lid.

2) None of the birds drink from a bottle with a lid colour whose first two letters match the first two letters of their name.

3) The milk the blackbird drinks contains twice as much fat as the milk the robin drinks.

Which birds drink which type of milk, with which lid colour?

....

OVERFISHING AND THE BARREN OCEANS

Industrialized fishing has taken an awful toll on the marine ecosystem and fished species. A 2014 study concluded that the biomass of predatory fish (which includes almost all the species that humans eat) has declined by two-thirds over the last hundred years, with most of this decline – 54 per cent – occurring in the last 40 years. A 2015 report by the World Wildlife Fund for Nature (WWF) said that populations of marine vertebrates including fish, turtles, birds, whales, dolphins and seals fell by half between 1970 and 2010. The same WWF study found that populations of tuna, mackerel and bonito dropped by 74 per cent in the same period. Some species, such as right whales, leatherback turtles and blue whales, have declined by 90 per cent or more.

The UN says that almost 90 per cent of the world's fisheries are fully fished or over-fished. Figures like these make less surprising, though no less disturbing, the assertion of the Ellen Macarthur Foundation that, by 2050, there will be more plastic than fish in the ocean. Some experts predict that this colossal depletion of biomass and top predator species will have catastrophic consequences for the global ecology in ways we don't yet fully understand – for instance, cutting off the circulation of nutrients between different ocean layers, which might, in turn, reduce the entire ocean to a watery desert.

PUZZLE INSTRUCTIONS

Place the fishing lines in the right directions so as to not overfish the sea areas. Draw an arrow representing a fishing line in each shaded box, pointing horizontally, vertically or diagonally. Every arrow must point to at least one number. The arrows must be placed so that each square has the given number of arrows pointing to it.

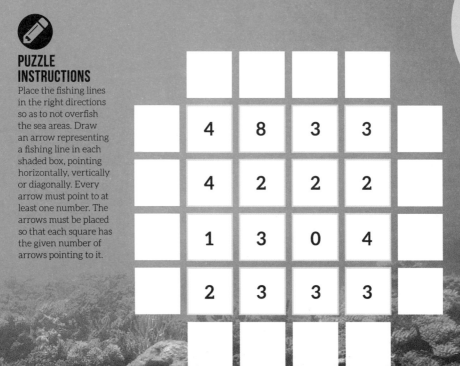

THE OLDEST TREES IN THE WORLD

The oldest known tree in the world was, until recently, Methuselah, a Great Basin bristlecone pine in the White Mountains in California which is over 4,850 years old. Its crown has now been taken by an unnamed near neighbour of the same species (*Pinus longaeva*), revealed in 2013 to be 5,062 years old. While there are many claims for greater longevity, these can be hard to assess because, in some ancient trees such as yews, older parts of the tree are lost and only younger ones retained. In other cases, the trees are clonal, which is to say that all or part of the tree is a clone of an older original. Thus, the Llangernyw Yew – an ancient tree in a churchyard in north Wales – is variously claimed to be 5,000 years old or "only" 1,500 years old. Conclusive dating has been impossible because the tree has grown out from a (now absent) core to leave a ring of "descendant" material.

Ancient clonal trees include Old Tjikko, a Norway spruce in Sweden with a root system dated at 9,000 years old (although its trunk is only around a century old), and a grove of quaking aspens in Utah, known as "Pando", which is collectively 40,000 years old, even though individual trees in the colony may only be decades old.

START

FINISH

PUZZLE INSTRUCTIONS
Navigate through the rings of the tree to get from one side to the other. Find your way from the entrance at the top through to the exit at the bottom.

THE PLANTS AND ANIMALS QUIZ

1) Which type of tree loses its leaves in the winter?

2) Which of the following is not technically a vegetable?

a) tomato ☐

b) broccoli ☐

c) potato ☐

d) celery ☐

3) What type of elephant has a trunk with a single tip, and small ears: Asian or African?

4) Which of the following is a flowering plant?

a) oak tree ☐

b) moss ☐

c) seaweed ☐

d) fir tree ☐

5) Which of the following animals would win a 100 m sprint: tuna, tiger, elephant?

6) Lemurs are native to which country?

7) Which nutritious element is extracted from the atmosphere and added to the soil by leguminous plants such as clover and alfalfa?

8) Snails, oysters and octopuses all belong to which phylum of animal?

9) Which three inputs do plants require in order to photosynthesize?

10) Which animal might you find on land in the Antarctic, the Galapagos and South Africa?

11) Which hydrocarbon provides plant cells with a tough cell wall?

12) Humans belong to which family of animals?

_____

13) What are the male parts of a flower called?

a) stamens ☐

b) carpels ☐

c) stigma ☐

14) Which of the following creatures is most closely related to the elephant?

a) shrew ☐

b) tapir ☐

c) aardvark ☐

d) manatee ☐

15) Which of the following is not a berry in the botanical sense?

a) banana ☐

b) tomato ☐

c) pomegranate ☐

d) strawberry ☐

e) kiwi ☐

16) What is the collective noun for a group of crows?

a) thief ☐

b) conspiracy ☐

c) murder ☐

17) By what name is the echidna better known?

_____

18) What is the common name given to the purple bell-shaped flowers of the digitalis?

_____

19) A male *Onthophagus taurus* can pull up to 1,140 times its own body weight, making it the world's strongest insect. What name is it better known by?

_____

20) What is the largest native predator among Australian land animals?

_____

PLANTS AND ANIMALS

SPACE

It has become passé to call space the "final frontier" – and there is still much to discover here on Earth – but it is undoubtedly true that, over the coming century, mankind will turn its eyes to the sky with increasing urgency. We can expect manned flights to Mars, permanent colonies in space, an increasing understanding of the vast mysteries of the universe and, perhaps, the definitive answer to the most exciting question of all... is there anyone else out there?

NEWTON'S CANNON

Many people imagine that once an astronaut rockets into space, gravity somehow cuts off and he or she ends up floating around the cabin in zero G. In fact, an astronaut in orbit around the Earth is not in zero G because the International Space Station (ISS) experiences 90 per cent of the force of the Earth's gravity. The reason the astronauts on the ISS are "weightless" and float around is that they are in free-fall, constantly falling toward the Earth, as if they were skydiving from an enormous height. The only reason they don't crash into Earth – but remain in orbit – is that they keep missing the ground!

Astronauts and other things in orbit are the realization of Isaac Newton's fantasy about a cannon. In a 1687 thought experiment, Newton pointed out that a projectile fired horizontally from a cannon follows a parabolic trajectory that brings it curving down to Earth. The faster the projectile is travelling when it leaves the cannon, the further it will get before it hits the ground. If the projectile is travelling fast enough, it will get all the way around the Earth and back to the cannon before hitting the ground. This is the principle on which orbital spaceflight is based. Astronauts, satellites and other objects are accelerated to great enough speeds such that they are always falling over the horizon; only when they slow down do they crash back to Earth.

The speed necessary for a cannonball to orbit the Earth is around 16,000 mph. The moon, on the other hand, is much smaller than the Earth and has much weaker gravity and no atmosphere. A high-velocity bullet with a muzzle velocity of around 1,200 metres per second would travel right around the moon... and hit the shooter in the back of the head.

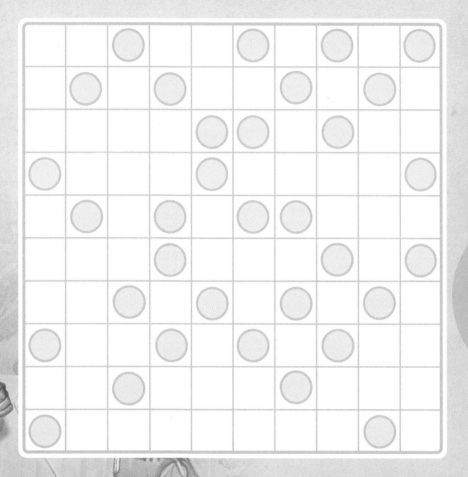

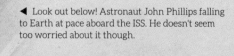

◀ Look out below! Astronaut John Phillips falling to Earth at pace aboard the ISS. He doesn't seem too worried about it though.

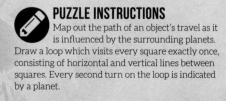

PUZZLE INSTRUCTIONS

Map out the path of an object's travel as it is influenced by the surrounding planets. Draw a loop which visits every square exactly once, consisting of horizontal and vertical lines between squares. Every second turn on the loop is indicated by a planet.

WHY ISN'T EARTH LIKE MARS OR VENUS?

Mars and Venus have much in common with Earth. All three are rocky planets that orbit in the right zone around their star for it to be possible for them to have liquid water on the surface. Indeed, scientists are pretty sure that water once flowed on the surface of Mars, and suspect that Venus, too, used to be much more hospitable than it is now.

Yet the climates of Mars and Venus have diverged radically from that of Earth, in opposite directions. Mars now has an extremely thin atmosphere and is cold and arid, while Venus has a crushingly thick atmosphere of toxic acid, with pressures 92 times higher than those on Earth. It has a surface hotter than Mercury, with temperatures high enough to melt lead.

Why are they so different from Earth? Mars is relatively small which gives it weaker gravity, so it was unable to hold on to the atmosphere it once had. Earth also has a metallic core that generates a magnetic field, which, in turn, protects it from the harsh radiation that has helped strip away the Martian atmosphere. Earth's plate tectonics ensure that carbon is constantly cycled between the ground and the atmosphere, maintaining a life-friendly blanket of greenhouse gases in the atmosphere. On Mars, carbon dioxide reacted with water and rock during its early history and, with no large-scale volcanic activity to recycle it into the atmosphere, the carbon was locked away forever.

Venus had the opposite problem. Most likely because it is just a little too close to the sun, water remained in its atmosphere as steam – a potent greenhouse gas – driving up temperatures, which, in turn, released masses of carbon from the rocks to trigger a runaway greenhouse effect, heating the surface to 477 °C.

◥ Mars has an average temperature of -63 °C.

◀ As well as being unbearably hot, Venus is notable for having a day that is longer than a year – 243 Earth days to 224 Earth days respectively.

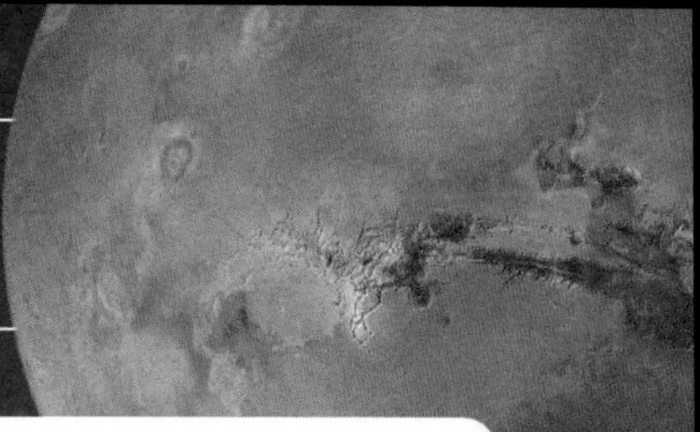

A 20×20 logic-puzzle grid with the following number clues (by approximate row and column):

					5				5	5									
				2	2	2		2	1				3						
		2	2	2				2				3			7				
		4	4							1		7	4	4		4			
				4			5			5									
			4	4		5		3			4	4		1			2		
	4				6		6		3							4	2		
2	2		4	5	6								11						
5		6			4		1											9	
					4									9				5	
			4				1						1	4		5	5		
				5	1											4			
5	3				7														11
							8	8	7										
	3	3														2	5		
								4								2			
		3	2			4	2								2	2			
		2	2	2			2						3		3				
			2	2			3				2	2	2						
					3	4			4	2									

PUZZLE INSTRUCTIONS

Coalesce the fragments to reveal a celestial picture. Draw paths to join numbers into pairs of identical numbers, where each path visits the number of squares equal to the numbers it joins, including those at either end. Paths can only move horizontally or vertically between squares, and no more than one path can enter any square. Once complete, shade all squares containing paths to reveal a hidden image.

THE SEARCH FOR EXOPLANETS

Until relatively recently, the idea of spotting planets orbiting stars – known as exoplanets – was the realm of science fiction, apart from in our own solar system, of course. But today nearly 4,000 exoplanets have been discovered, while another 3,000 are strongly suspected.

Since the early 1990s, when exoplanet discoveries began, the number of known exoplanets has doubled approximately every 27 months. There are two main ways in which astronomers can spot an exoplanet. The first is by measuring the minute characteristic wobble in a star's rotation, which is caused by the gravitational influence of an orbiting planet. The second is by detecting the minuscule dimming of light we observe from a star when a planet passes – or transits – between us and the star. Incredibly, scientists can use these tiny observations to work out orbital distances and planetary masses. This data, in turn, gives strong clues about a planet's composition and likely surface conditions. Now, it is even possible to analyse the light that passes through exoplanetary atmospheres to obtain direct clues about their chemical make-up.

PUZZLE INSTRUCTIONS

Work out the size of the various planets based on the measurements given. Place 1 to 6 once each into every row and column of the grid. Place digits in the grid in such a way that each given clue number outside the grid represents the number of digits that are "visible" from that point, looking along that clue's row or column. A digit is visible unless there is a higher digit preceding it, reading in order along that row or column. For example, in "216435" the 2 and 6 are visible from the left, but 1 is obscured by the 2, and the rest by the 6.

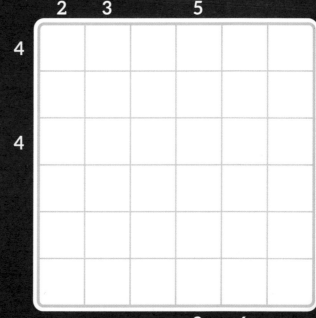

WE ARE ALL STARDUST

With the exception of some products of radioactive decay, nuclear explosions and high-energy particle collisions in accelerators or with cosmic rays, every atom that makes up the Earth, and everything on or in it, is much older than the Earth or solar system. The simplest atoms, such as hydrogen, helium and some lithium, may be over 13 billion years old, since they were created in the "Big Bang". Everything else, from the oxygen you breathe to the gold ring you wear on your finger, was created by a process known as nucleosynthesis.

This is where the nuclei of lighter elements are crushed together with such fantastic force that they merge to create heavier nuclei. These heavier nuclei, in turn, are combined until all of the elements of the periodic table are created. Nucleosynthesis of elements up to iron in terms of atomic weight can happen in ordinary stars (which can be bigger than our own sun); elements heavier than iron are only created in supernovae and collisions between neutron stars. The atoms that make up the Earth coalesced from a disc of dust and gas, which also formed the sun and other planets. The disc itself was made of the remnants of previous stars that had exploded and dissipated. Therefore, all of our atoms were originally dust left over from other stars.

PUZZLE INSTRUCTIONS

With the heaviest element at the top, work out how the given particles break down into components. Complete this number pyramid by writing a number in each empty brick, so that each brick contains a value equal to the sum of the two bricks immediately beneath it.

SPACE

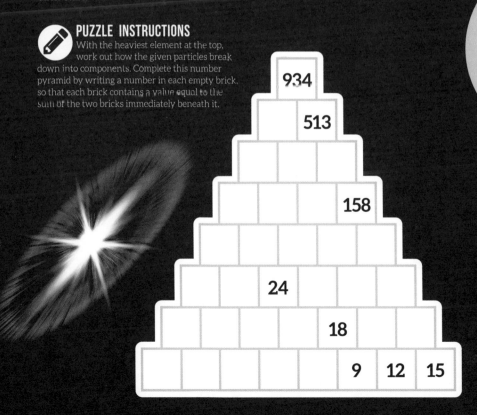

MISSION TO MARS

SPACE

PUZZLE INSTRUCTIONS

Find the water sources from the instrument readings given. Place water sources into some empty squares in the grid. Clues in some squares show the number of water sources in touching squares – including diagonally. No more than one water source may be placed per square.

NASA has committed to a programme of space exploration that will culminate in a manned mission to Mars, although SpaceX founder and private-rocket magnate Elon Musk may well get there first, as his company is developing a huge rocket with that aim in mind.

That said, getting to Mars is a massive undertaking, which poses challenges we are not yet able to overcome. Staying there will be even harder. The first challenge is one of distance; even when Earth and Mars are closest together, the journey to Mars would still take roughly 6–8 months (although Musk claims his rocket could do it in 80 days, and may be even quicker in the future). This will be a psychological endurance test for a small crew in close quarters. More importantly, however, it will be physically hazardous, as low gravity weakens the muscles and bones, and the astronauts will be bombarded with potentially lethal doses of radiation from cosmic rays. To set up a base on Mars, humans will need to take almost everything they need to survive with them. Hopefully, Martian ice can be tapped for water and fuel, while Martian regolith (the stony soil) can be used to construct at least the outer layers of shelters, to protect habitations from the extreme cold and the deadly radiation. Establishing long-term colonies on Mars will require new technologies to cope with this radiation, compensate for the low gravity (just 38 per cent of that on Earth), and gather enough sunlight and other resources to grow food.

	2			1		3	
3		2		2	3		
			4		5		3
						4	2
2		4	4				1
	1			5			1
		2				4	
	1	1	1		3		

SPACE

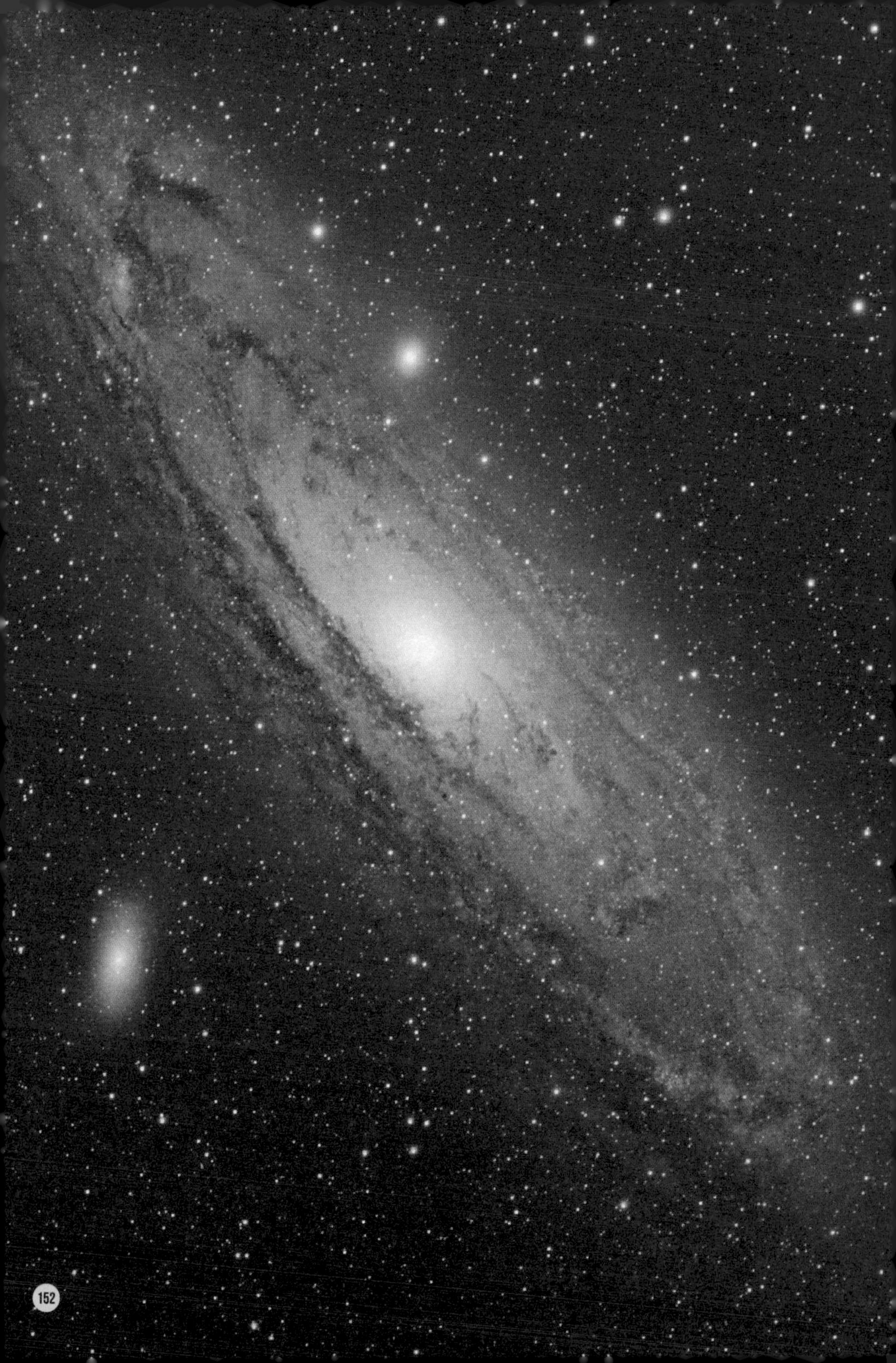

THE SIZE OF THE UNIVERSE

Distances in space are... well, astronomical, and to make describing them more manageable, astronomers use units much bigger than miles or kilometres. Since the speed of light in a vacuum is a constant, it makes a useful basis for measuring – in the form of light years. A light year is, quite simply, the distance travelled by light in a year. Since light in the vacuum of space travels extremely fast (almost 300,000 kilometres per second), in the space of a year, it travels almost 9.5 trillion km (about 5.9 trillion miles).

The nearest star to Earth – Proxima Centauri – is over 40 trillion km away, and the nearest spiral galaxy outside our own Milky Way – the Andromeda Galaxy – is about 2.4×10^{19} km distant. Describing these distances with light years is much less unwieldy: Proxima Centauri is 4.25 light years away and the Andromeda Galaxy is 2.5 million light years away. Modern estimates put the diameter of the observable universe at around 93 billion light years.

◀ Our friendly neighbour, the Andromeda Galaxy.

SPACE

PUZZLE INSTRUCTIONS

Plot the various spiral galaxies on the starmap. Draw along some of the dashed grid lines in order to divide the grid up into a set of regions. Every region must contain exactly one circle, and the region must be symmetrical in such a way that if rotated 180 degrees around the circle it would look exactly the same.

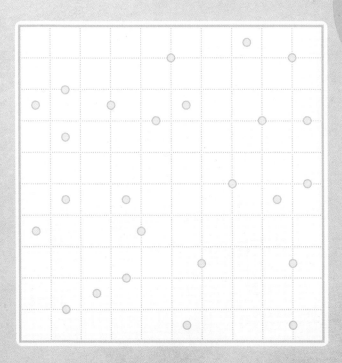

INGREDIENTS FOR A UNIVERSE

Although scientists have discovered hundreds of sub-atomic particles, and can peer back through time to the "Big Bang", the nature of the vast majority of the universe remains a complete mystery. We know that all of the matter and energy we can account for comprises just 5 per cent of the universe. We can tell, for example, from the way that galaxies are rotating, and from the rate of expansion of the universe, that there must be forms of matter and energy about which we know nothing. Because this matter and energy seem not to be affected by any of the electromagnetic forces we rely upon for detector technology – and, hence, cannot be "seen" by us – scientists call them dark matter and dark energy. To account for the speed of galactic rotation and the acceleration of cosmic inflation, it is reckoned that about 68 per cent of the universe is dark energy, 27 per cent is dark matter, and just 5 per cent is the stuff we can observe and that we know anything about.

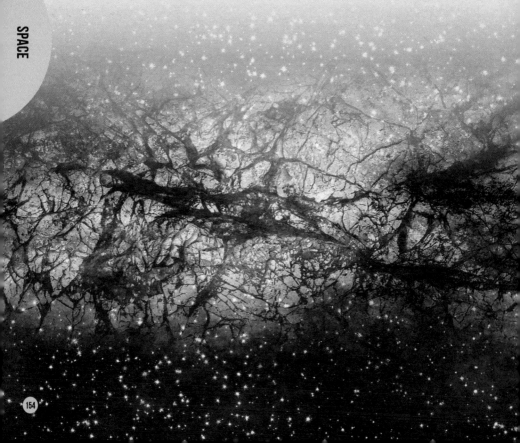

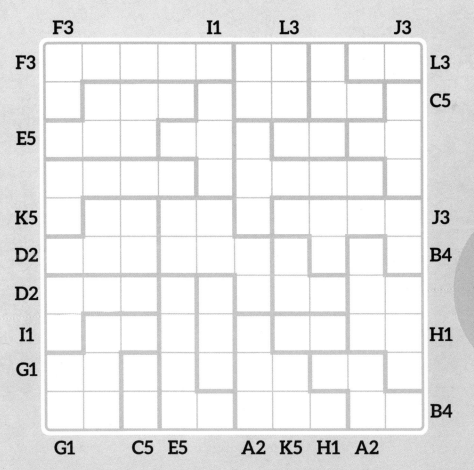

F3   I1   L3   J3

F3 .... L3
.... C5
E5
.... J3
K5
D2 .... B4
D2
I1 .... H1
G1
.... B4

G1   C5 E5   A2 K5 H1 A2

SPACE

PUZZLE INSTRUCTIONS

Work out where the real matter in the form of mirrors are, bearing in mind that most squares are filled with dark matter and have no effect, appearing empty. Draw diagonal lines across certain squares to form mirrors, with exactly one mirror per bold-lined region. The mirrors must be placed so that a laser fired into the grid from each lettered clue would then exit the grid at the same letter elsewhere, having bounced off the exact number of mirrors indicated by the number next to the letter.

SPACE

In the 1920s, the American astronomer Edwin Hubble was able to use a trick of the light to measure the velocity of distant galaxies. His extraordinary conclusion was that not only are the galaxies hurtling through space at incredible speeds, but also that all of them are moving away from each other. In other words, the universe is expanding. The same phenomenon was implied by Einstein's equations of general relativity, and working through the calculations had led the Belgian priest and physicist Georges Lemaitre to follow this to its logical conclusion: if the universe is expanding, it must have started off at a single point.

Lemaitre referred to this theoretical starting point of the universe as the "hypothesis of the primeval atom", or the "Cosmic Egg". The now-familiar term, the "Big Bang", would not be coined until many years later by British astronomer Fred Hoyle. Lemaitre described the Cosmic Egg as "exploding at the moment of the creation". He would also poetically refer to the beginning of time and space as a "now without yesterday".

▶ The Hubble space telescope was launched in 1990 and was the first major optical telescope in space.

▼ Fred Hoyle, the man who coined the term the "Big Bang" – although he said it with disdain. He died in 2001 never having accepted that the Big Bang ever happened.

PUZZLE INSTRUCTIONS

Colour the interior of the spiral to reveal a hidden space object. It's best to start at the outside, otherwise you may accidentally colour the areas outside the spiral.

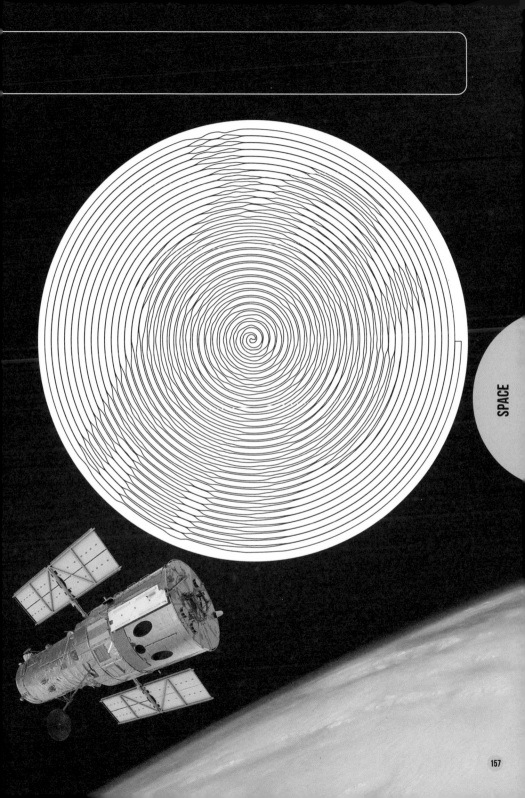

BLACK-HOLE MYSTERIES

A black hole is a place where gravity has become so overwhelming that it pulls matter and energy into a point of infinite density, known as a "singularity". Because the gravitational attraction of the singularity is so immense, it pulls in anything that comes closer than a threshold known as the "event horizon". Because no light can escape, reflect off, or pass through the event horizon, it is utterly black – hence "black hole".

To the layperson, black holes are confusing and mysterious. What would it be like to fall into one? Does matter that falls into a black hole come out somewhere else? But cosmologists also ponder black-hole mysteries. The work of Stephen Hawking and others has shown that the event horizon of a black hole can radiate particles (known as "Hawking radiation"), which will eventually cause the hole to evaporate. However, this raises the paradox that information which has fallen into the black hole will disappear when the hole disappears, apparently violating the laws of conservation. Cosmologists are still trying to formulate theories that work around this.

PUZZLE INSTRUCTIONS

Find your way to the centre of the black hole from the entrance at the top.

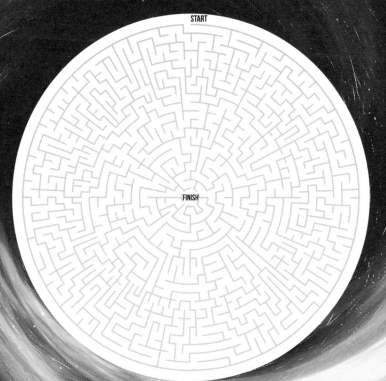

START

FINISH

THE FATE OF THE UNIVERSE

We are now pretty sure that the universe started off with a "Big Bang", but how will it end? Will it end?!

Initially, cosmologists thought that the determining factor would be the density of matter in the universe versus the rate at which the universe is expanding. If the gravitational attraction between the matter were greater than the expansionary force, the universe would eventually contract in a "Big Crunch". But if the expansionary force were greater, the universe would continue to expand forever. If the two were perfectly balanced, the expansion would slow to zero at some point infinitely in the future.

However, now that we know that most of the universe is made of dark energy, the picture gets a lot more complicated. The density of dark energy seems to remain constant as the universe expands but, since we don't know what it is or how it works, it is hard to predict what will happen next. If dark energy increases the expansionary pressure on the universe, it could lead to a "Big Rip" scenario, in which all matter – from galaxies and stars to atoms and sub-atomic particles – is eventually ripped apart!

PUZZLE INSTRUCTIONS

Some space is completely unknown, and perhaps unknowable. Place a digit from 1 to 9 into each empty square, so that no digit repeats in any row, column or bold-lined 3×3 box. Shaded squares should not have numbers added, and may represent different missing numbers in each of their row, column and 3×3 box regions than that which you might otherwise expect.

7	2		1	5				
		6			4	9	7	
			2			5	3	
		5		8	7			
								1
9		8		2		6		3
						7	6	
3	4	1			8			
				7			1	

THE SPACE QUIZ

SPACE

1) Which is hotter: the surface of Venus or the surface of Mercury?

2) Which of the following is NOT a gas giant?

a) Jupiter ☐

b) Neptune ☐

c) Pluto ☐

3) Which space telescope had its faulty eyesight fixed by a Space Shuttle mission in 1993?

4) Which of the following can you NOT see from the International Space Station with the naked eye?

a) the Great Wall of China ☐

b) the Grand Canyon ☐

c) the Great Barrier Reef ☐

5) Where in the solar system would you find the Great Red Spot?

6) Neil Armstrong and Buzz Aldrin were the first men on the moon. Who was the third member of the *Apollo 11* mission?

7) Which of the solar system's planets are known to have rings?

8) Which of these stars is closest to Earth?

a) Sirius ☐

b) Proxima Centauri ☐

c) Barnard's Star ☐

9) Which is the most common element in the universe?

10) What name is given to the threshold beyond which nothing can escape the gravity of a black hole?

11) Where in the solar system would you find Olympus Mons, Mare Erythraeum, Hellas Basin and Valles Marineris?

12) Which type of astronomical body is believed to originate from the Kuiper belt?

13) Who discovered Uranus?

14) Where would you find the Oort cloud?

15) What planet-destroying event took place in 2006?

16) If both were dropped from a height of 2 metres, which would land first: an apple dropped on Mars or one dropped on Earth?

17) In which constellation is the star Betelgeuse?

18) The Crab Nebula is a vast cloud of dust and gas left over from what cosmic event?

19) Where would you find Fotla Corona, Baltis Vallis and Ishtar Terra?

20) A pulsar is a rotating, highly magnetized astronomical object that emits EM-radiation, but what type of object is it?

SPACE

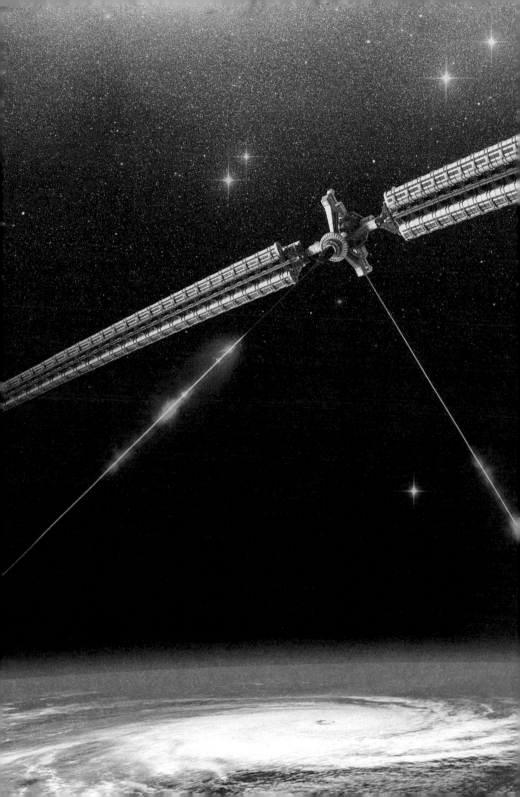

TECHNOLOGY

For millennia, the pace of human technological change was negligible; it mainly consisted of fashioning stone and animal parts into tools. Over the past few centuries, however, the speed of development has increased almost exponentially, with the advent of the computer age promising ever greater future advancements. Whether it leads to some sort of *Black Mirror*-esque dystopian hellscape, or merely a new Apple gadget every month, technological change is certain to shape the 21st century in ways we can't even imagine yet.

THE SIX SIMPLE MACHINES

▼ Wheels are as obviously crucial today as they ever have been. The earliest depictions of wheeled vehicles are 8,000 years old in what is modern-day Syria.

▶ Wedges have been used all over the world, from prehistoric hand axes to modern-day inidigenous antler wedges.

A machine is a device for doing work – in the technical sense where "work" means applying force to move something. A simple machine is something with a single or no moving parts that makes work easier by concentrating (in one fashion or another) force applied over a larger distance into force applied over a shorter distance. This magnifies, or "multiplies", the force, which is why such a device is known as a force multiplier.

The ancient Greeks recognized six basic types of simple machine: the wheel, the lever, the inclined plane, the pulley, the screw and the wedge. In fact, the latter three are versions or combinations of the former three; a screw, for instance, is an inclined plane wound in a spiral.

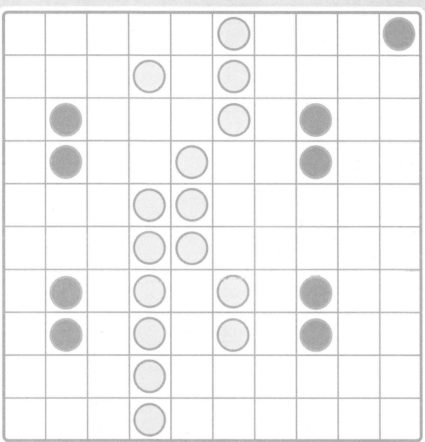

PUZZLE INSTRUCTIONS

Create a pulley loop that turns and goes straight around or over various rollers. Draw a single loop, using only horizontal and vertical lines, that passes through the centre of every circle. The loop must pass straight through every white circle without turning, and then make a 90-degree turn in either or both of the previous and following squares. Conversely, the loop must make a 90-degree turn at every pink circle, but then travel straight through the previous and following squares without turning. The loop cannot visit any square more than once.

DYSON SPHERES

A Dyson sphere is a hypothetical futuristic construct proposed by English-American theoretical physicist, mathematician and visionary Freeman Dyson. Dyson suggested that an advanced civilization seeking to harvest all the energy radiated by its sun might surround the star with solar-power gathering surfaces, eventually building up a complete shell or sphere around the sun. This, in turn, suggests a way to look for advanced extrasolar civilizations, since such a device might affect how a distant star appears to us, in a telltale way.

A stellar-radiation profile featuring only dim infrared light might be the signature of such a sphere and, thus, proof of intelligent life elsewhere in the universe.

▲ Freeman Dyson's thought experiments were legendary. He also came up with the idea of a tree that could grow on a comet and help sustain life there, which is now a classic sci-fi trope.

▶ This is the kind of thing we are looking for – just much bigger!

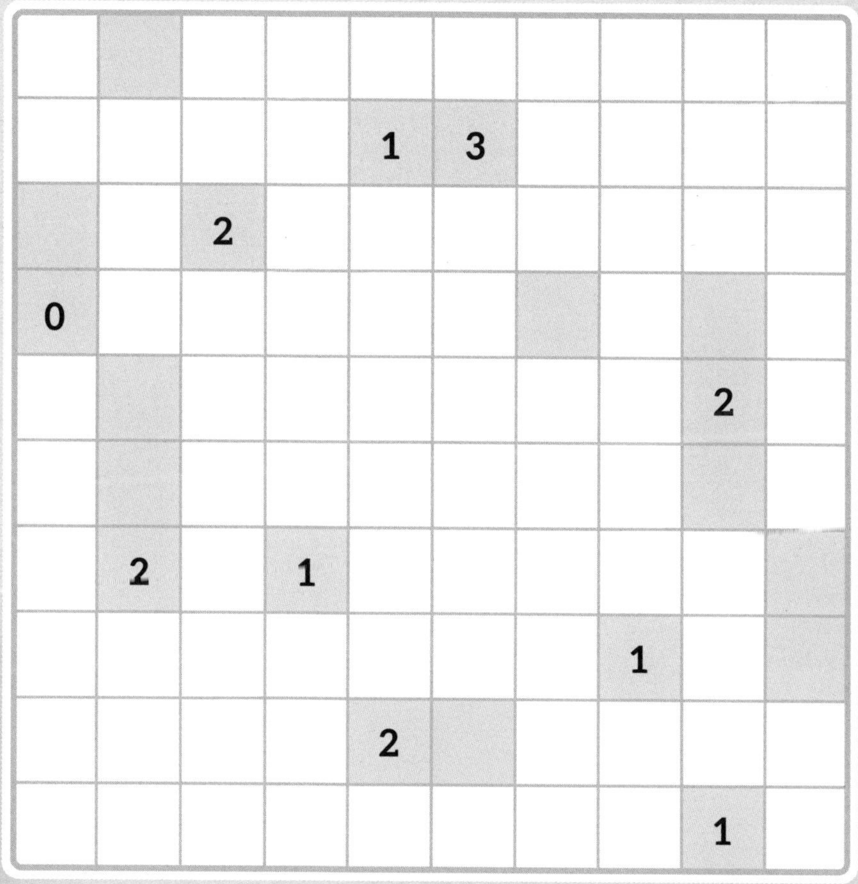

PUZZLE INSTRUCTIONS

Work out how best to place the light bulbs to maximise the light shone around the grid, based on the clues. Place light bulbs in some empty cells so that each empty cell is illuminated. Light bulbs illuminate every cell in their row and column until blocked by a pink cell. No light bulb can be illuminated by another light bulb. Some pink cells have numbers, which correspond to the number of light bulbs in their adjacent, non-diagonally touching, cells. Not all light bulbs are necessarily clued. All pink cells are given.

THE SPACE ELEVATOR

The main limiting factor for getting an object into orbit stems from the amount of fuel a rocket must carry in order to lift its own weight. Frustratingly, of course, the more fuel it has to carry, the heavier it gets. Circumventing this challenge would dramatically cut the expense and difficulty of getting into orbit. One futuristic suggestion is for an elevator that can climb a cable from the Earth to orbital altitude.

Such a cable could be kept taut by a counterweight orbiting at its far end, in a geostationary orbit. You can imagine this by thinking of whirling around a bucket of water on a string. If the counterweight orbited at a distance of around 95,000 km, a large platform could be constructed part of the way up the cable, in low Earth orbit, at roughly the altitude of the International Space Station. The anchor point of the cable would be at the equator. The main challenge facing such an undertaking is that there currently exists no material strong enough to cope with the colossal tension involved, yet light enough not to collapse under its own weight.

PUZZLE INSTRUCTIONS

Construct a series of different height elevator towers. Place 1 to 7 once each into every row and column of the grid. Place digits in the grid in such a way that each given clue number outside the grid represents the number of digits that are 'visible' from that point, looking along that clue's row or column. A digit is visible unless there is a higher digit preceding it, reading in order along that row or column. For example, in '214756' the 2, 4 and 7 are visible from the left, but 1 is obscured by the 2, and the rest by the 7.

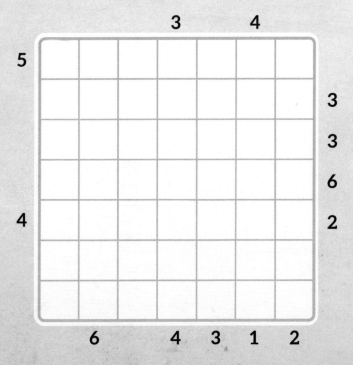

TECHNOLOGY

THE SECRET OF CEMENT

Cement is a mortar that can be used to stick together bricks and blocks, or mixed with gravel to create concrete, a kind of artificial rock. The Romans used volcanic ash to create a cement they called *pozzolan*, which could even be used underwater. The secret of waterproof cement was lost for many centuries, but was rediscovered in the late Middle Ages, leading eventually to the invention of Portland cement by Joseph Aspidin in 1824. His cement could also set underwater, as graphically proven when a ship carrying barrels of cement powder sank off the coast of England. The barrels of now-solid cement were rescued and used to build a pub.

PUZZLE INSTRUCTIONS

Fit the blocks together to form both a perfect fit and a perfect pattern. Place the given pieces once each into the grid so that all squares are filled. You can rotate them but not reflect them. They must be placed in such a way that the coloured squares form a second complete set of the exact same pieces as below, e.g. the four orange squares join to create another piece. No two such created pieces of the same colour can touch.

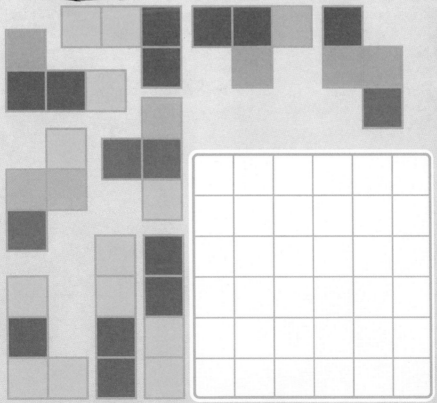

DIODES AND TRANSISTORS

Computers work by using logic gates to perform operations on bits of information. The fundamental requirement for a computer is therefore to be a device that can represent a binary bit. In other words, it must be a device that can contain information that can be either "on" or "off"; or "open" or "closed". An electronic version of such a device – a valve that can allow or block the transmission of electricity – was discovered by British electrical engineer John Fleming in 1904, when he created the first diode valve. It was a diode because it had two electrodes in it; later the valve was improved to a triode. These valves were similar to light bulbs, which meant they were bulky, power-hungry and unreliable.

The first computers of the 1940s, built with valves, were huge room-filling affairs. The development of semi-conductor materials led to the creation, in 1947, of the transistor – essentially a low-power, lightweight, solid-state triode valve that could be miniaturized, ushering in the age of micro-electronics.

 ◢ An example of Fleming's electrode valve from 1904.

PUZZLE

Each logic operation must be applied between a pair of lit segments (i.e. the white segments that make up each number), as follows:

- & = logical AND: only keep segments lit if they are lit in both the inputs
- + = logical OR: light all segments that are lit in EITHER or both inputs
- x = logical EXCLUSIVE OR: light segments only if they are lit in one of the inputs, but not if lit in both of them

Here's an example, showing the output, so you can see how it works:

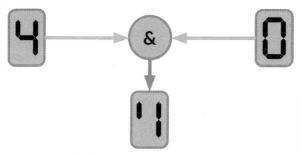

Notice how only the segments in common between the 4 and 0 are kept in the output, due to the logical AND.

PUZZLE INSTRUCTIONS

Decode the message by applying the logic gate transformations to the input data. What are the logic gates concealing? Follow the arrows and apply the logical operations to find out!

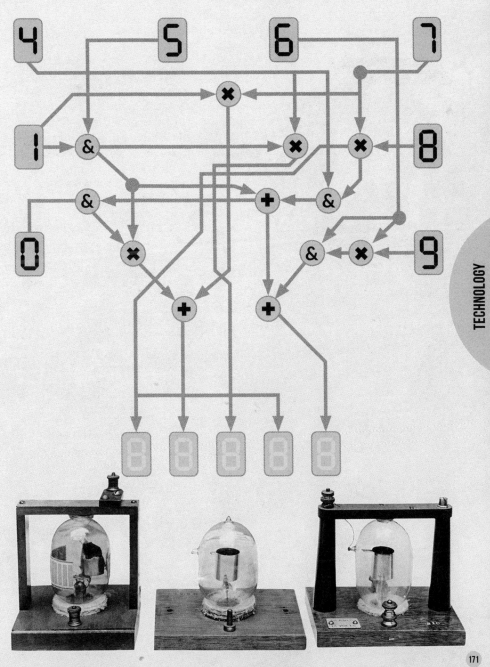

THE TURING TEST

English mathematician Alan Turing was a pioneer in the theory of computers. His work proving that computing machines could solve all sorts of complex problems prompted speculation about intelligent – or thinking – machines, and Turing was often asked whether he believed such a thing might be possible. He dismissed the question as meaningless, insisting that all that can be asked is whether a machine *appears* intelligent. Turing said that if a human, in conversation with a machine, could not distinguish that machine from a human intelligence then, to all intents and purposes, the machine would have to be considered intelligent. He suggested a variation on a popular parlour pastime, known as the "imitation game", in which judges tried to guess the gender of unseen writers from typed responses to questions. A similar format is today applied to computer programs conversing with human judges, who try to guess whether they are real or artificial.

▲ Alan Turing: war hero, computer genius, and finally getting the recognition he deserves.

▶ It might have the looks, but does it have the brains?

PUZZLE INSTRUCTIONS

Discover if the binary image conceals a robot or a human. Shade some squares according to the clues at the start of each row or column. The clues provide, in reading order, the length of every run of consecutive shaded squares in each row and column. There must be a gap of at least one empty square between each run of shaded squares in the same row or column.

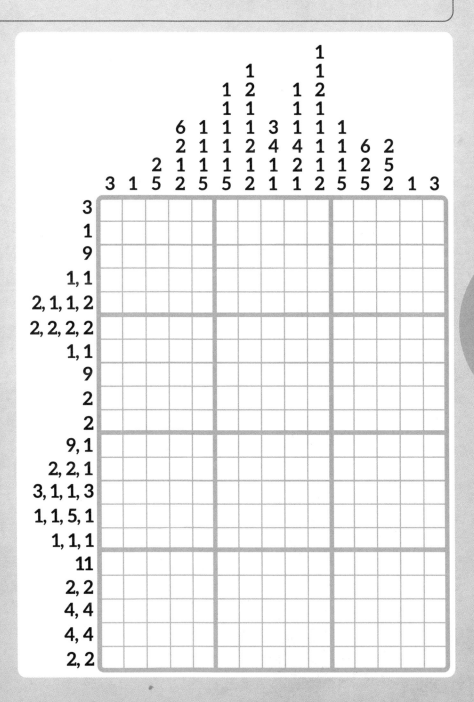

TECHNOLOGY

HOW MUCH DOES AN E-MAIL WEIGH?

TECHNOLOGY

Thanks to Einstein's famous equation – $e=mc^2$ – we know that energy has mass, albeit a very tiny amount. This means that electronic representations of things – such as computer programs, picture files, and the Internet itself – all have mass. Calculations suggest that the entire Internet weighs about 50 g – roughly as much as a large strawberry – but these sums are complex.

Let's start with something easier, like an e-mail. The first thing to know is that the physical representation of a bit of information in a computer is in the form of an electrical charge in a device called a capacitor. If the bit is a 1, the capacitor is charged with about 40,000 electrons. A typical e-mail contains about 50 kilobytes of information. There are 8 bits in a byte and 1,024 bytes in a kilobyte so, in total, a typical e-mail is made up of 409,600 bits. However, only the bits that are 1s actually involve a charge; 0s involve no electrons. On average, about half of the bits will be 1s, so the total number of electrons making up the e-mail is 204,800 X 40,000 ~ 8 billion electrons. An electron does not weigh very much: 9.109×10^{-31} kg, or roughly 10^{-27} g. This gives us a calculation for the weight of the e-mail of 8 billion × 10^{-27} g = 8×10^{-18} g. An e-mail therefore weighs about 8 quintillionths of a gram, or 0.000000000000000008 g.

▼ It has fewer cat GIFs than the Internet.

PUZZLE INSTRUCTIONS

Work out the weights of the various objects using the balances. The pictures show three perfectly balanced scales, with four different types of shape. Each shape has a different weight, and the lightest shape weighs 2 kg. Given this, how much does each of the four types of shape weigh?

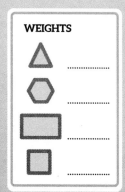

WEIGHTS

BIOMIMETICS

Biomimetics is the practice of copying – or at least taking inspiration from – nature. There are numerous examples of natural engineering that far outperform anything humans can yet achieve. The human heart, for instance, can beat up to 4 billion times or more before it wears out or breaks down. Avian wings are lighter, stronger and more flexible than any technological equivalent. Evolution has honed near-perfect natural designs over billions of years through countless iterations.

Accordingly, an emerging trend in engineering is to take advantage of this process by copying its best products. A classic example is the invention of Velcro, the microscopic hook-and-loop fastener; its creator George de Mestral was inspired by the way that plant burrs clung to his dog's fur.

◀ Velcro is integral to life in space. It is used to attach objects to walls, fasten spacesuits and even for flaps through which astronauts can scratch their noses.

PUZZLE INSTRUCTIONS

Join each velcro hook (pink circle) to a fibre (white circle). Draw horizontal or vertical lines to join the circles into pairs, so each pair contains one pink and one white circle. Lines cannot cross either another line or a circle.

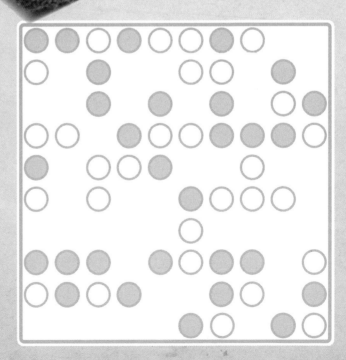

NUCLEAR FUSION

Nuclear fusion is the process that powers the sun, and it has long been the dream of engineers to harness it on Earth in order to provide almost limitless, low-cost, pollution-free energy. Fusion power is generally thought of as involving the merging of two hydrogen nuclei to form a helium nucleus. Its only product would be an inert gas, and only minute quantities of material would need to be involved in the production process.

The challenge is that achieving fusion requires extremely high temperatures. The primary technology being developed to achieve fusion is a toroidal (donut-shaped) ring, in which powerful magnets are used to confine and concentrate hot gas, which then needs to be heated to 150 million °C – 10 times hotter than the core of the sun. But almost every aspect of the design is extremely complex and very expensive, and even if it works, it will only be a demonstration device. Other designs involve high-power X-ray lasers and ultra-high-power guns that fire pellets together with colossal force.

PUZZLE INSTRUCTIONS

Join the pairs of particles (numbers) into a single joined pair. Draw a series of separate paths, each connecting a pair of identical numbers. No more than one path can enter any square, and paths can only travel horizontally or vertically between squares.

1	2	3					4		
					2			3	
			5						
6			7			8			
1							9		
						4	8		
				7					
				6				9	
						5			

HYPERLOOP

In 2012, the tech entrepreneur Elon Musk outlined his vision for a new mode of transport that would overcome some of the problems that limit the speed of high-speed trains, specifically air resistance and rolling resistance (friction involving wheels). The concept involves sealed tubes at low or zero air pressure, within which would run pressurized passenger pods, propelled along by electromagnetic linear induction motors, while being lifted off the rails or floor by a system called maglev.

Maglev uses magnets to create a cushion of air between the train and the track. Another set of magnets is used to propel the train forward. With little or no air or rolling resistance to overcome, the pod can be accelerated to high speeds – around 760 mph (1,200 km/h) – and then coast along with minimal energy input. Musk described the idea as a "cross between Concorde and a railgun and an air-hockey table". His team of engineers published their initial plans and made the project open source. Since then, several companies and teams have competed to develop test tubes, prototype motors and feasibility studies for full-scale systems.

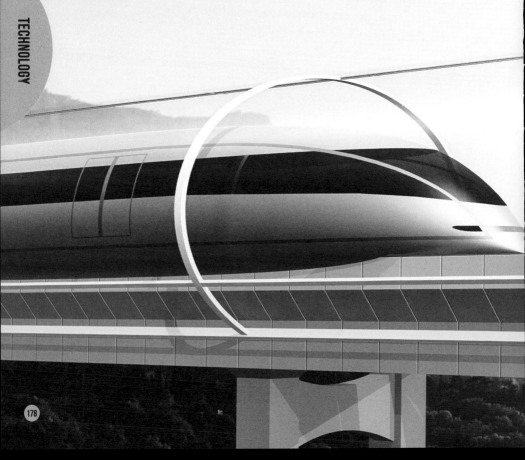

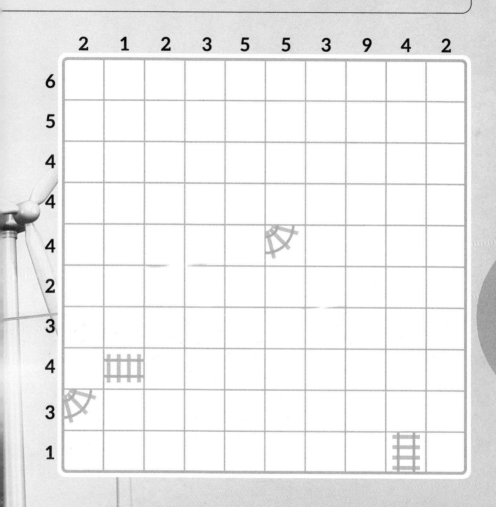

	2	1	2	3	5	5	3	9	4	2
6										
5										
4										
4										
4										
2										
3										
4										
3										
1										

PUZZLE INSTRUCTIONS

Work out the route of the maglev railway. Draw railway pieces in some squares in order to complete a track that travels all the way from its entrance in the leftmost column to its exit in the bottom row. It can't exit the grid or cross over itself. Figures outside the grid reveal the number of railway pieces in each row and column.

SELF-DRIVING CARS

Truly autonomous cars – aka self-driving, or robot cars – are the subject of intensive research and development. They could revolutionize motoring and transport, but also wider concepts of ownership, life-work patterns and mobility for disadvantaged groups. In order for this to happen, however, self-driving cars need to master several types of challenge, including perception of the environment, path planning, car control and navigation.

The first of these poses the greatest challenge because, even when equipped with sensors like infrared, ultrasound, radar and lidar, machine intelligences can still struggle to spot hazards.

Another area of challenge is what is known as the trolley problem. At some point, an autonomous driving agent will be faced with a situation where it has to choose between two bad options – for example, hitting a pedestrian or swerving and crashing into a wall. How should it make this choice, and who will be held responsible for it? The passenger, the owner, the programmers, or the car manufacturers?

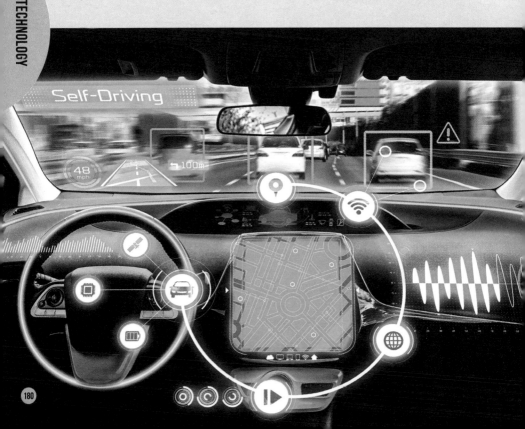

4				4			6		
			7						
	7								
									4
5						6			
				4		6	4		
			3						
	7		3					7	
				1					

PUZZLE INSTRUCTIONS

Plot a loop route based on the readings from the beacons. Draw a single loop that passes through some of the empty squares, using horizontal and vertical lines. The loop cannot reenter any square. The loop must pass through the given number of touching squares next to each number clue, including diagonally touching squares.

CAN VENICE BE SAVED?

As demonstrated by extreme flooding in late 2019, Venice is under severe threat from rising sea levels, especially since the whole city has actually sunk considerably over the years. The sediments on which it was built have naturally compacted over centuries, but the sinking problem was considerably worsened in the twentieth century by extraction of groundwater from an aquifer beneath the city. Although that was halted, the city had already sunk by 25 cm (10 in), and continues to sink because of tectonic shifting.

While climate change threatens to make the regular *acqua alta* (high water) events catastrophic, the city will soon be under the protection of a giant flood-defence system named the MOSE Project. When it is eventually working, mobile gates will rise up to block the three inlets between the sea and the Venetian lagoon whenever a storm surge or high tide threatens. Unfortunately, corruption and mismanagement have delayed completion of the project, which is now forecast to be in operation in 2022.

PUZZLE INSTRUCTIONS

Work out the island shapes that are above sea level. Shade some empty squares so that each number in the grid becomes part of an unshaded area of precisely that many squares. There must be exactly one number in each unshaded area. Shaded areas cannot cover a 2×2 or larger square area. All shaded squares must form one continuous area. Areas are not considered continuous if they only touch diagonally.

▼ The Grand Canal in Venice, with the Rialto Bridge in the background.

		2					5	
	3					3		
	5			2				
					3		5	
		6					4	
	1					3		

TECHNOLOGY

THE TECHNOLOGY QUIZ

1) What kind of engine was pioneered by Frank Whittle in the 1930s?

2) What name is given to the positive electrode in an electrolytic cell?

3) What does FM stand for in radio?

4) What is the meaning of the acronyms AC and DC?

5) What lifesaving piece of safety apparatus was invented by Humphry Davy in 1815?

6) Digging started in 1881 on which great civil-engineering project that would not be formally opened until 1994?

7) The suffix "rtf" in a computer file name stands for what?

8) Where might you find woofers and tweeters?

9) By what name is the vehicle initially known as a landship now better known?

10) What type of pump is named after Archimedes?

11) What is unusual about a Wankel engine?

12) What was the name of the first electronic digital computer, developed by the British in 1943?

13) Which simple machine multiplies force by rotating around a pivot, allowing the points of application and distribution of force to be separated?

14) What kind of current does a polyphase brushless motor run on?

15) What force keeps airplanes aloft?

16) Which phenomenon did Michael Faraday elucidate with his work involving magnets, wires and batteries?

17) What is the element most commonly used in batteries for applications such as mobile phones and electric cars?

18) A Stirling engine is powered by what sort of energy?

19) What name is given to an interconnected series of real or virtual artificial neurons (units that mimic the action of nerve cells) used for computation?

20) Which of the following is NOT a type of logic gate used in electronic circuits?

a) NAND ☐

b) XAND ☐

c) OR ☐

d) NOT ☐

TECHNOLOGY

SOLUTIONS

🧠 THE BRAIN

Page 11

PHINEAS GAGE: THE MAN WHO HAD A ROD THROUGH HIS BRAIN

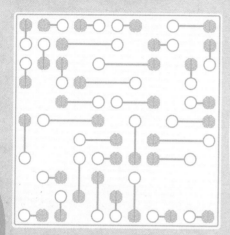

Page 15

THE BEAST WITHIN

Page 13

MIND OVER MATTER

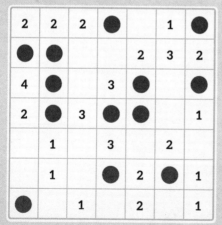

Page 17

THE BRAIN OLYMPICS

3 ③	5 ⑤	2 ④	③6	4 ④	1	⑧8 ③	9 ②	7
9 ⑨	1 ⑥	⑧8 ②	7	5	3	4 ④	②2	③6
7	6	4 ②	8	9	2 ②	1 ②	③3	5 ⑤
6 ⑥	8 ④	5	3	7	9	2	④4 ②	④1 ⑨
1 ④	②2	3	④4 ③	6	5	7	8	9 ③
4 ②	7	9 ⑨	①1	②2	④4	8	5	6 ②
8 ④	3 ③	②6	③2	②1	①7 ③	7	9	5 ⑤	4 ②
2	9	7 ⑦	5	3	4	6 ⑥	①1 ⑦	⑧8 ②
5	4 ④	①1	⑨9	8	6 ②	3	7	2

SOLUTIONS

Page 18
MARS VS VENUS

V	M	M	V	V	M	M	V
V	M	V	M	M	V	V	M
M	V	M	V	M	M	V	V
M	V	V	M	V	M	M	V
V	M	V	M	M	V	V	M
M	V	M	V	V	M	V	M
M	M	V	M	V	V	M	V
V	V	M	V	M	V	M	M

Page 20
THE MAN WHO MISTOOK HIS WIFE FOR A HAT

Page 19
NEURAL NETWORKS IN MAN AND MACHINE

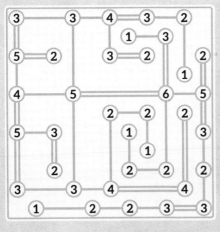

Page 22
NO LIMITS

The four items that were replaced are:

Cup & saucer; **Hairbrush**; **Plant pot**; **Batteries**

And they have been replaced by:

Bowl; **Comb**; **Different plant pot**; **Torch**

Page 23
WHY DO WE DREAM?

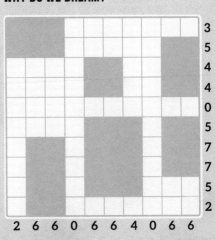

Page 25
THE INVISIBLE GORILLA

It is possible to see both a vase in the centre of the image, or two faces looking at each other on either side of the image.

Page 26
THE BRAIN QUIZ

1. How many hemispheres does the human brain have? **Answer:** Two

2. What is the name of the outermost part of the human brain? **Answer:** c)

3. Is brain size correlated with intelligence? **Answer:** Yes, although only weakly.

4. What is the name given to the phenomenon where a person feels like they have seen something before? **Answer:** déjà vu

5. Roughly how many neurons does a human brain contain? **Answer:** b)

6. What is the branch of psychology that studies thinking, memory and language? **Answer:** c)

7. What does REM stand for? **Answer:** Rapid Eye Movement

8. What were the three parts of Sigmund Freud's tripartite theory of personality? **Answer:** id, ego and superego

9. Which of the following did Freud NOT class as a phallic symbol? **Answer:** It's a trick question – Freud claimed that ALL of them were phallic symbols.

10. The Clever Hans phenomenon is where cues unconsciously exchanged between participants trigger conditioned responses. What was Clever Hans? **Answer:** a horse

11. Which of these brain disorders involves movement and co-ordination? **Answer:** a)

12. Three of the most common phobias in the UK are *brontophobia* (fear of thunder), *cardiophobia* (fear of the heart and heart conditions) and *aeronausiphobia* (fear of air sickness). What is the most common phobia? **Answer:** *arachnophobia* (fear of spiders)

13. Koro is an example of a type of psychogenic illness known as a culture-bound syndrome, which involves fear of magical influence causing shrinkage or disappearance of what type of organ? **Answer:** genital

14. What delusions characterize the condition known as Fregoli syndrome? **Answer:** c) – Fregoli syndrome is named after a famous quick-change artist of the late nineteenth and early twentieth centuries; a) is the characteristic delusion of Capgras syndrome; b) is the characteristic belief of the Mignon delusion

15. What is the Baskerville effect? **Answer:** a)

16. What part of the nervous system causes blushing and goosebumps? **Answer:** the autonomic nervous system – to be precise, the sympathetic branch of the autonomic nervous system

17. What is the name given to the phenomenon whereby a patient is cured of illness by a sugar pill believed to be a potent medicine? **Answer:** the placebo effect

18. What is the name given to the phenomenon whereby a person is made ill by what they believe to be a poison, even though it is harmless and inactive? **Answer:** the nocebo effect

19. Which lobe of the brain is associated with vision? **Answer:** the occipital lobe

20. By what popular name is *parapraxis* better known? **Answer:** Freudian slip

CHEMISTRY

Page 30
THE QUEST FOR THE PHILOSOPHER'S STONE

	7					1
		8		3		
	7				4	
5			6	5	9	
1						
		7	5		5	8
	2				3	9

Page 32
DECODING SCIENTIFIC NOTATION

A=9; **B**=5; **C**=6; **D**=14

Page 33
THE ANGRY BEE IN THE SHRINKING BOX

Page 34
THE PERIODIC TABLE

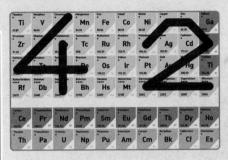

Page 36
RADIOACTIVITY

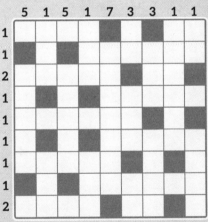

Page 37
HOW MANY BARRELS OF PISS DOES IT TAKE TO DISCOVER A NEW ELEMENT?

11	11	5	6	6	6	7	7
11	11	5	5	6	7	7	7
11	11	5	2	6	6	7	7
11	3	5	2	8	8	8	4
11	3	8	8	8	3	3	4
11	3	8	6	6	6	3	4
11	5	8	5	6	6	6	4
11	5	5	5	3	3	3	1

Page 41
WHAT IS A CHEMIST'S FAVOURITE ANIMAL?

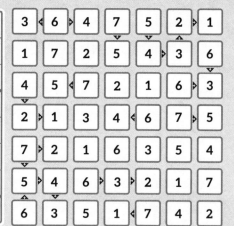

Page 39
AN EXPLOSIVE DISCOVERY

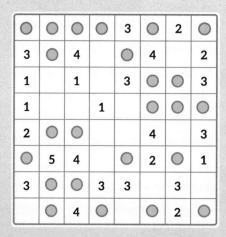

Page 42
BUCKYBALLS AND NANOTUBES

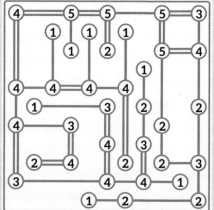

Page 43
WATER IS WEIRD

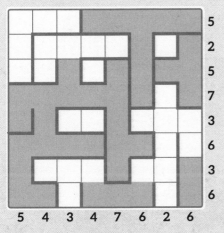

Page 46
KEKULÉ'S DREAM

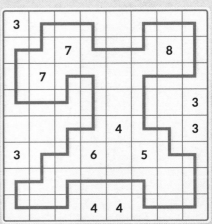

Page 45
HOW TO BUILD A BETTER BATTERY

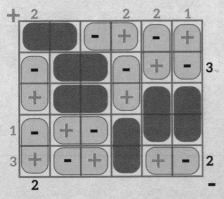

Page 47
FINGERPRINTS OF LIGHT

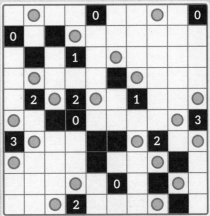

Page 49
THE DEADLIEST POISON

Page 50
THE CHEMISTRY QUIZ

1. What is the chemical formula of water?
   Answer: H_2O

2. What do you get if you burn hydrogen in air?
   Answer: water vapour and, potentially, a big explosion

3. With which element did the alchemists associate the sun and urine? **Answer:** gold

4. Which compound of two elements gives rust and blood their red colour? **Answer:** iron oxide

5. What element is the lead in a pencil made from? Clue: it's not lead. **Answer:** carbon – pencil lead is actually graphite, a form of carbon

6. If you measured the pH level of a liquid and found that it was 1, would it be an acid or an alkali? **Answer:** acid – a pH of 1 is extremely acidic

7. What is the smallest and lightest element in the periodic table? **Answer:** hydrogen

8. What is the name given to an atom that has lost or gained electrons and so has an electric charge? **Answer:** an ion

9. What is the densest element? **Answer:** c)

10. The atomic number of oxygen is 8. How many protons does an atom of oxygen have? **Answer:** 8 – atomic number is the number of protons an element contains

11. What is the collective name for the group of elements that includes neon, xenon, krypton and argon? **Answer:** noble gases

12. Which metal can dissolve gold? **Answer:** mercury

13. What is the name given to a cloud, or soup, of ions? **Answer:** plasma

14. What elements make up a molecule of ethanol? **Answer:** carbon, oxygen and hydrogen

15. Saltpetre is one of the key ingredients of gunpowder. What is the chemical name of saltpetre? **Answer:** a)

16. Which French scientist was the first to discover evidence of radioactivity? **Answer:** b)

17. The Haber Process is fundamental to the production of synthetic fertilizer. It combines nitrogen with hydrogen to produce which chemical? **Answer:** ammonia

18. What is the weight of a mole of carbon-12? **Answer:** 12g – a mole of an isotope weighs the same as its atomic mass

19. Which isotope of uranium is fissile (i.e. can be used to engender nuclear fission)? **Answer:** b)

20. Sodium has an atomic number of 11. What comes after it in the periodic table? **Answer:** c) – magnesium has atomic number 12

THE ENVIRONMENT

Page 54
THE GREENHOUSE EFFECT

8	3	7	5	2	9	4	1	6
4	9	6	3	7	1	2	5	8
1	5	2	4	6	8	7	3	9
6	8	4	7	1	3	5	9	2
9	7	1	2	4	5	6	8	3
3	2	5	9	8	6	1	4	7
7	4	9	8	5	2	3	6	1
2	6	3	1	9	4	8	7	5
5	1	8	6	3	7	9	2	4

Page 59
ICE AGES

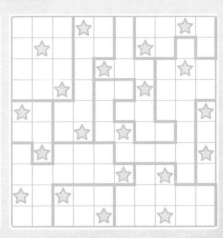

Page 56
GEOENGINEERING

	4		4		3
	1	4	2	5	3
	2	5	3	1	4
	3	1	4	2	5
	5	3	1	4	2	3
2	4	2	5	3	1

3

Page 61
RENEWABLE ENERGY

8	1	2	7	4	3	9	5	6
7	4	9	6	5	2	8	1	3
3	7	4	2	9	1	6	8	5
1	5	7	8	3	6	4	2	9
4	2	6	9	7	8	5	3	1
9	3	5	1	8	4	2	6	7
6	8	3	5	1	9	7	4	2
2	9	8	3	6	5	1	7	4
5	6	1	4	2	7	3	9	8

Page 57
TIPPING POINTS

76	5	19
16	83	1
8	12	80

Page 63
JOURNEY TO THE CENTRE OF THE EARTH

Page 65
HOT AND WET

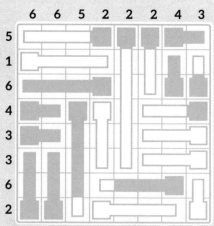

Page 64
HURRICANES

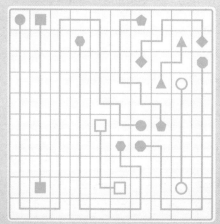

Page 67
FALLING TO EARTH

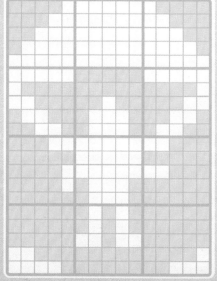

Page 69
FIRE SEASON

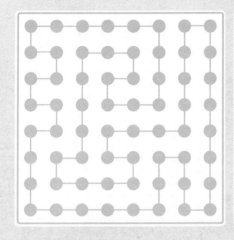

Page 71
MEGA-TSUNAMIS

1	2	1	4	2	3	1	2
4	1	3	2	1	4	2	3
1	3	2	1	5	2	3	1
2	1	4	3	1	5	1	2
1	2	5	1	3	1	4	1
3	1	2	4	1	2	5	3
2	4	3	1	2	1	3	2
4	3	1	2	1	3	2	4

Page 70
PREDICTING EARTHQUAKES

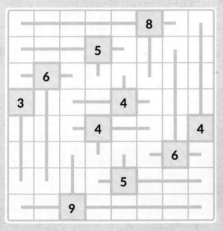

Page 73
THE NEXT SUPER-VOLCANO

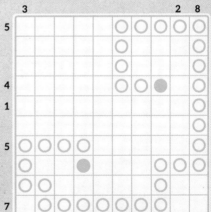

SOLUTIONS

197

THE ENVIRONMENT QUIZ

1. At which pole would you find Antarctica? **Answer:** the South Pole

2. Which natural phenomenon is also known as a twister? **Answer:** a tornado

3. Which is the biggest continent? **Answer:** Asia

4. What is the biggest river in the world by volume of discharge? **Answer:** the Amazon

5. The Amazon rainforest spreads across nine countries, including Brazil, Suriname, Peru, Guyana and French Guiana. Can you name any of the other four countries? **Answer:** Ecuador, Venezuela, Colombia or Bolivia

6. What is the name given to the flow of water from the land to the rivers to the sea and then back again via evaporation, cloud formation and precipitation? **Answer:** the water cycle

7. Name the five ocean basins recognized by the US Board on Geographic Names **Answer:** Atlantic, Pacific, Indian, Arctic and Southern (Antarctic)

8. Which is the layer of the atmosphere in which humans live? **Answer:** the troposphere

9. What source of energy is being tapped in a geothermal power plant? **Answer:** heat from the Earth's tectonics

10. In which direction does a hurricane spin? **Answer:** anti-clockwise – hurricanes are also known as anticyclones; in the southern hemisphere such storms spin clockwise, but they are not called hurricanes

11. What is the name of the layer of the Earth beneath the crust? **Answer:** the mantle

12. What is the lowest point on the Earth's surface? **Answer:** the Challenger Deep in the Marianas Trench at the bottom of the Pacific

13. Which is the most potent greenhouse gas? **Answer:** c)

14. An earthquake registering 6 on the Richter Scale is how many times more powerful than one registering 5? **Answer:** 10 times more powerful – the Richter scale is logarithmic

15. What is the name given to the global cooling that can occur after a huge volcanic eruption? **Answer:** volcanic winter

16. What is the opposite of a neap tide? **Answer:** a spring tide

17. What name is given to the periods between Ice Ages? **Answer:** c)

18. What is the name of the scale used to rate the impact hazard of asteroids and comets? **Answer:** b)

19. Roughly how many trees are there in the world? **Answer:** c)

20. What is the name given to the climate cycle in the Pacific Ocean, which has a global impact on weather patterns, and which occurs every few years? **Answer:** El Niño

SOLUTIONS

HEALTH

Page 78
SYNTHETIC DNA

	T	A		C	
	T	A	C	G		G
A	A	G		T	C	C
C		C	G	A	T	
	G		T	C	A	A
C	C	T	A		G	
			C	G	

Page 82
CRISPR GENE EDITING

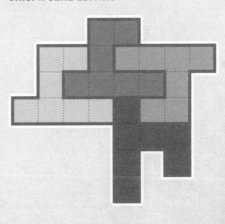

Page 81
BIONIC PROSTHETICS

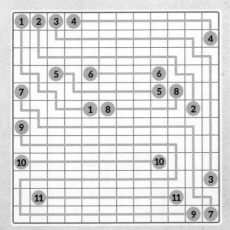

Page 83
ANTIBODIES AND THE IMMUNE SYSTEM

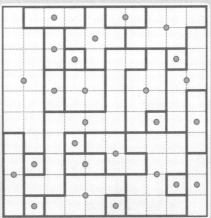

SOLUTIONS

Page 85
HERD IMMUNITY

Page 87
DEATH MAP

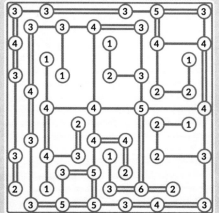

Page 86
WORLD'S WORST DISEASES

Page 89
HEAD TRANSPLANTS

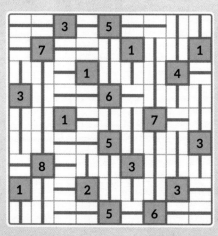

Page 91
IS 10,000 STEPS ENOUGH?

Page 93
ANTIBIOTIC RESISTANCE

V	V	A	V	A	V	A	A
A	V	V	A	V	A	V	A
V	A	A	V	A	A	V	V
V	V	A	V	A	V	A	A
A	A	V	A	V	A	V	V
V	V	A	V	A	A	V	A
A	A	V	A	V	V	A	V
A	A	V	A	V	V	A	V

Page 94
THE HEALTH QUIZ

1. Who discovered the antibiotic properties of penicillin? **Answer:** a)
2. Which part of the eye perceives light and generates nerve signals to send via the optic nerve? **Answer:** the retina
3. Which blood type is known as the "universal donor"? **Answer:** blood type O
4. From which organ does the aorta transport blood? **Answer:** the heart
5. Is influenza caused by a bacterium or virus? **Answer:** a virus
6. Which disease is believed to have caused the Black Death in the Middle Ages? **Answer:** bubonic plague
7. What colour of blood cell mediates the human immune response? **Answer:** white
8. Which is the human body's biggest organ? **Answer:** the skin
9. How many teeth does an adult human have? **Answer:** 32
10. What does the acronym ECG stand for? **Answer:** electrocardiogram
11. Which fire-retardant fibrous material can cause mesothelioma cancer in the lungs? **Answer:** asbestos
12. How many chromosomes are there in a typical human cell nucleus? **Answer:** 46
13. What is the biggest bone in the human body? **Answer:** femur, or thigh bone
14. Against which diseases does the MMR vaccination protect? **Answer:** mumps, measles and rubella
15. What are the letters that make up a DNA sequence? **Answer:** G, A, C and T – they stand for the bases guanine, adenine, cytosine and thymine
16. Which toxic metal was used as a treatment for syphilis in the Middle Ages? **Answer:** mercury
17. Roughly how many times does the human heart beat in an average lifetime? **Answer:** a)
18. Roughly how many bacteria live in your intestines? **Answer:** c)
19. Which organ of the body makes insulin? **Answer:** pancreas
20. Where would you find renal calculi? **Answer:** the kidneys – renal calculi are kidney stones

SOLUTIONS

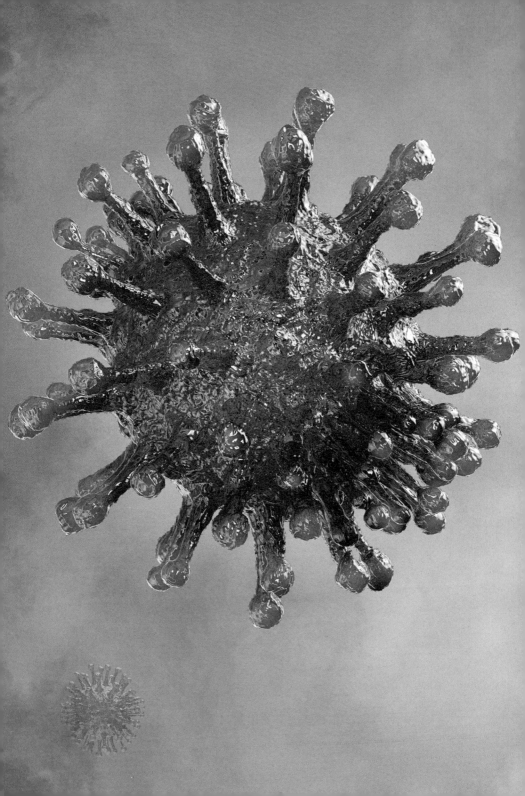

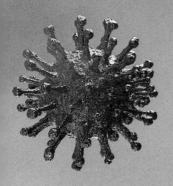
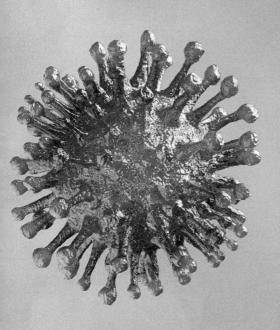

⚛ PHYSICS

Page 98
LOOK OUT BELOW

	1			
3	5	2	4	1
5	1	3	2	4
4	2	1	5	3
1	4	5	3	2
2	3	4	1	5

2 (left, row 3), 4 (left, row 5), 2 (right, row 2), 3 (right, row 4)
3 (bottom, col 1), 3 (bottom, col 4)

Page 101
THE SLIT EXPERIMENT

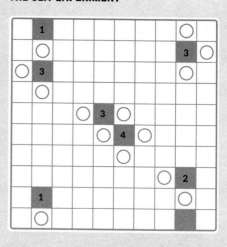

Page 99
SCHRÖDINGER'S CAT

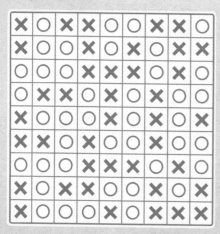

Page 103
TRAVELLING AT THE SPEED OF LIGHT

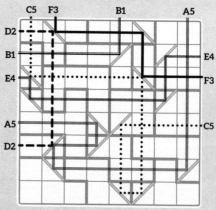

SOLUTIONS

Page 104
SPLITTING THE ATOM

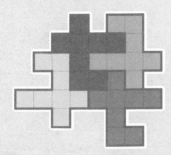

Page 105
STRING THEORY

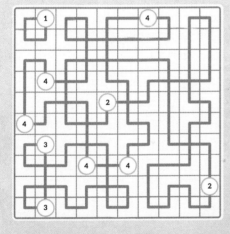

Page 108
RAINBOWS AND BLUE SKIES

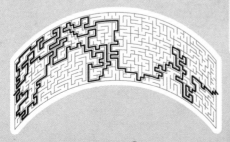

Page 107
NEWTON'S APPLE

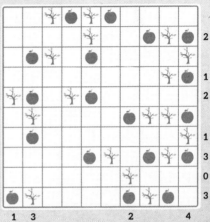

Page 109
THE COLDEST PLACE IN THE UNIVERSE

61	58	59	56	55	46	45	44
62	60	57	54	49	39	47	43
63	64	53	50	38	48	40	42
20	52	51	29	30	37	36	41
21	19	28	1	12	31	33	35
22	27	18	13	2	11	32	34
23	26	14	17	3	6	10	9
25	24	16	15	5	4	7	8

Page 111
THE BIGGEST SCIENTIFIC EXPERIMENT IN HISTORY

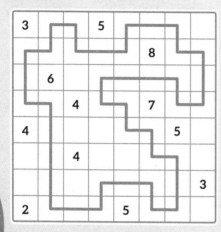

Page 114
FRIDGE MAGNETS VS THE EARTH

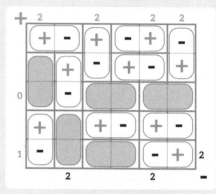

Page 115
TIME-TRAVELLING TWINS

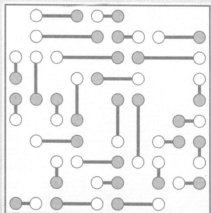

Page 113
SPOOKY ACTION AT A DISTANCE

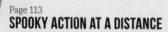

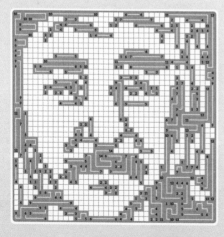

Page 117
THE WEIGHT OF AIR

1	7	2	8	3	9	5	4	6
8	3	5	4	2	6	7	1	9
4	9	6	1	5	7	3	2	8
9	6	3	7	1	2	8	5	4
2	5	8	6	4	3	9	7	1
7	1	4	9	8	5	2	6	3
6	2	7	3	9	1	4	8	5
5	4	9	2	6	8	1	3	7
3	8	1	5	7	4	6	9	2

Page 118
THE PHYSICS QUIZ

1. Which will fall faster in a vacuum: a feather or a cannonball? **Answer:** both will fall at the same rate

2. What type of particle carries electromagnetic force? **Answer:** photon

3. Three containers are filled with boiling water. One is made of copper, one is made of cork and the third is made of ceramic. Which container will be the hottest to hold? **Answer:** the copper one – copper is a very good conductor of heat; better than ceramic or cork

4. Which part of the spectrum has a slightly longer wavelength than red light? **Answer:** infra-red

5. What type of energy is given to a ball by raising it to the top of a hill? **Answer:** c)

6. A high-powered rifle fixed 2 m off the ground fires a bullet horizontal to the ground. At the exact same instant that the bullet leaves the muzzle of the gun travelling at 160 m/s, the shell-casing drops from the rifle. Which will hit the ground first? **Answer:** they will both hit the ground at the same time

7. In the nucleus of an atom, which sub-atomic particle carries a positive charge? **Answer:** proton

8. For the detection of which phenomenon did the LIGO detector team win the Nobel Prize in Physics for 2017? **Answer:** gravity waves

9. What property is possessed by silicon and germanium, which makes them useful in electronics? **Answer:** they are semi-conductors

10. If you painted a wall with paint that absorbs light in the blue and red regions of the visible spectrum, what colour would it be? **Answer:** green

11. Which British physicist proved the existence of the electron in 1897? **Answer:** J.J. Thomson

12. What name is given to a machine that travels by burning fuel, which it carries to generate exhaust that is expelled in one direction to produce a force in the opposite direction? **Answer:** a rocket

13. Force equals mass multiplied by which quantity? **Answer:** acceleration

14. If you fill a bath to the brim and drop into it a cube of gold measuring 5 cm on each edge, how much water will slop over the edge of the bath? **Answer:** 25 cm^3 – the volume of the gold cube is 5^3 cm^3, which means that it will displace that amount of water

15. In Einstein's famous equation e=mc^2, what does "c" stand for? **Answer:** the speed of light in a vacuum

16. What does LHC stand for? **Answer:** c)

17. Spaceship A is twice as far away from the Earth as Spaceship B. How many times less will be the force of gravity it experiences? **Answer:** four – gravity decreases by the square of the distance

18. Which phenomenon causes the change in pitch of a siren that you hear when an ambulance speeds past? **Answer:** the Doppler effect

19. What is alpha radiation? **Answer:** a)

20. The giant Super-Kamiokande detector in Japan is looking for which particles? **Answer:** neutrinos

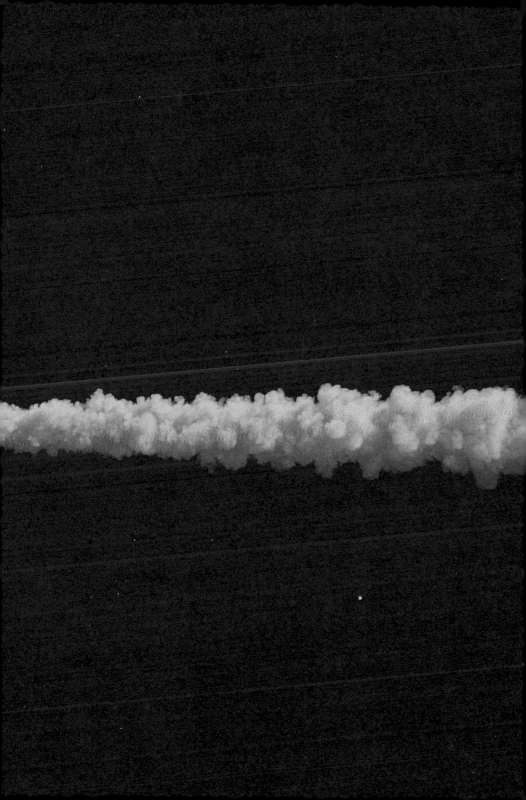

Page 123
CRYPTOZOOLOGY

```
F K E K A N A N S I E O M T A
S Q N O G A R D A N G S O S A
C A R E M I H C P P A Y K O H
S R R F W U Y M C S E P O N A
H B T B A K A R Q T T I L O R
N M N R A C E U I H T N E K A
I L D A O C A L U O G F M L W
H Y U P N T A N P R U I B A A
H H P M C D D P I I T B E I K
G I E H E E I F U S E I M B O
H E A I R R F B U H E G B H N
S U N B W I M M E H C F E P I
C G I I N O A A A A P O K A M
N R B O E D Y S I C R O E W T
D O E A E D I U R D N T M N O
```

Page 126
DARWIN'S MONSTER

Page 125
HOW MANY SPECIES ARE THERE ON EARTH?

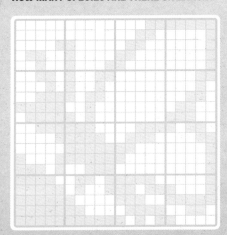

Page 127
EDGE CASE

5	5	2	2	3	3	2	7	7	7
5	5	5	7	7	3	2	7	7	7
7	7	7	7	2	2	3	3	7	8
8	8	7	8	8	8	1	3	8	8
1	8	8	8	2	2	8	8	8	8
5	5	5	5	3	3	8	5	5	5
7	7	7	5	7	3	7	7	2	5
7	1	4	4	7	7	7	1	2	5
7	7	4	4	1	7	3	3	3	3
7	9	9	9	9	9	9	9	9	9

Page 129
POISONOUS PLANTS

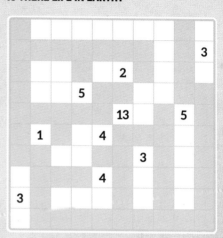

Page 132
IS THERE LIFE IN EARTH?

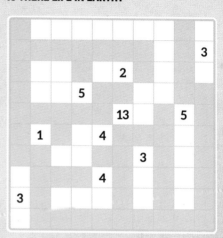

Page 131
THE INDESTRUCTIBLE WATER BEAR

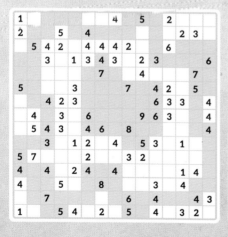

Page 133
ECHOLOCATION

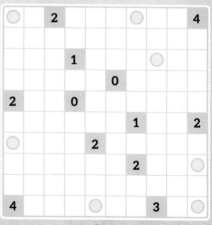

Page 135
WHO KILLED THE MAMMOTH?

Page 137
OVERFISHING AND THE BARREN OCEANS

Page 136
CLEVER CROWS AND PERSPICACIOUS PARROTS

The blue tit drinks the 1% milk with a red lid.

The robin drinks the 2% milk with a blue lid.

The blackbird drinks the 4% milk with a silver lid.

Page 139
THE OLDEST TREES IN THE WORLD

THE PLANTS AND ANIMALS QUIZ

1. Which type of tree loses its leaves in the winter? **Answer:** deciduous

2. Which of the following is not technically a vegetable? **Answer:** a)

3. What type of elephant has a trunk with a single tip, and small ears: Asian or African? **Answer:** Asian

4. Which of the following is a flowering plant? **Answer:** a)

5. Which of the following animals would win a 100 m sprint: tuna, tiger, elephant? **Answer:** tuna

6. Lemurs are native to which country? **Answer:** Madagascar

7. Which nutritious element is extracted from the atmosphere and added to the soil by leguminous plants such as clover and alfalfa? **Answer:** nitrogen

8. Snails, oysters and octopuses all belong to which phylum of animal? **Answer:** molluscs

9. In order to photosynthesize, plants require which three inputs? **Answer:** sunlight, water and carbon dioxide

10. Which animal might you find on land in the Antarctic, the Galapagos and South Africa? **Answer:** a penguin

11. Which hydrocarbon provides plant cells with a tough cell wall? **Answer:** cellulose

12. Humans belong to which family of animals? **Answer:** great apes

13. What are the male parts of a flower called? **Answer:** a)

14. Which of the following creatures is most closely related to the elephant? **Answer:** d)

15. Which of the following is not a berry in the botanical sense? **Answer:** d)

16. What is the collective noun for a group of crows? **Answer:** c)

17. By what name is the echidna better known? **Answer:** the spiny anteater

18. What is the common name given to the purple bell-shaped flowers of the digitalis? **Answer:** foxglove

19. A male *Onthophagus taurus* can pull up to 1,140 times its own body weight, making it the world's strongest insect. What name is it better known by? **Answer:** a dung beetle

20. What is the largest native predator among Australian land animals? **Answer:** the dingo

SOLUTIONS

213

SPACE

Page 145
NEWTON'S CANNON

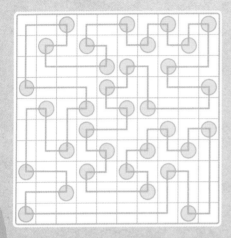

Page 148
THE SEARCH FOR EXOPLANETS

	2	3		5	
4	2	3	4	1	6	5
	6	4	1	2	5	3
4	1	2	5	3	4	6
	4	6	2	5	3	1
	5	1	3	6	2	4
	3	5	6	4	1	2
			2	6		

Page 147
WHY ISN'T EARTH LIKE MARS OR VENUS?

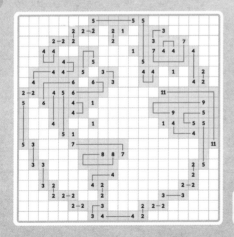

Page 149
WE ARE ALL STARDUST

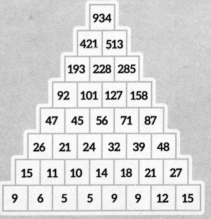

SOLUTIONS

Page 151
MISSION TO MARS

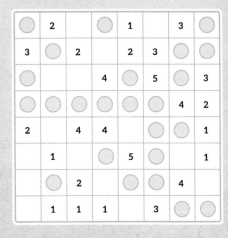

Page 155
INGREDIENTS FOR A UNIVERSE

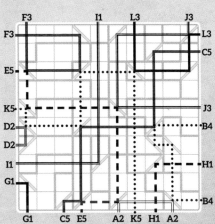

Page 153
THE SIZE OF THE UNIVERSE

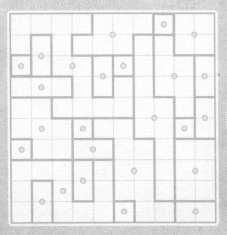

Page 157
THE BIG BANG

Page 158
BLACK-HOLE MYSTERIES

Page 159
THE FATE OF THE UNIVERSE

7	2	9	1	5	3		4	6
5	3	6	8		4	9	7	2
	1	4	2	9	6	5	3	8
1		5	6	8	7	4	2	9
2	6	3	9	4	5	8		1
9	7	8		2	1	6	5	3
8	5	2	3	1		7	6	4
3	4	1	5	6	8	2	9	
6	9		4	7	2	3	1	5

Page 160
THE SPACE QUIZ

1. Which is hotter: the surface of Venus or the surface of Mercury? **Answer:** the surface of Venus

2. Which of the following is NOT a gas giant? **Answer:** c)

3. Which space telescope had its faulty eyesight fixed by a Space Shuttle mission in 1993? **Answer:** the Hubble Space Telescope

4. Which of the following can you NOT see from the International Space Station with the naked eye? **Answer:** a)

5. Where in the solar system would you find the Great Red Spot? **Answer:** Jupiter

6. Neil Armstrong and Buzz Aldrin were the first men on the moon. Who was the third member of the *Apollo 11* mission? **Answer:** Michael Collins

7. Which of the solar system's planets are known to have rings? **Answer:** Saturn, Jupiter, Neptune and Uranus

8. Which of these stars is closest to Earth? **Answer:** b)

9. Which is the most common element in the universe? **Answer:** hydrogen

10. What name is given to the threshold beyond which nothing can escape the gravity of a black hole? **Answer:** the event horizon

11. Where in the solar system would you find Olympus Mons, Mare Erythraeum, Hellas Basin and Valles Marineris? **Answer:** Mars

12. Which type of astronomical body is believed to originate from the Kuiper belt? **Answer:** comet

13. Who discovered Uranus? **Answer:** William Herschel

14. Where would you find the Oort cloud? **Answer:** in a sphere around our sun, beyond the orbit of Pluto

15. What planet-destroying event took place in 2006? **Answer:** Pluto was downgraded from "planet" to "dwarf planet" by the International Astronomical Union

16. If both were dropped from a height of 2 metres, which would land first: an apple dropped on Mars or one dropped on Earth? **Answer:** the one on Earth – gravity is stronger on Earth, so falling objects accelerate more quickly

17. In which constellation is the star Betelgeuse? **Answer:** Orion

18. The Crab Nebula is a vast cloud of dust and gas left over from what cosmic event? **Answer:** a supernova

19. Where would you find Fotla Corona, Baltis Vallis and Ishtar Terra? **Answer:** Venus – they are all surface features

20. A pulsar is a rotating, highly magnetized astronomical object that emits EM-radiation, but what type of object is it? **Answer:** a neutron star

TECHNOLOGY

Page 165
THE SIX SIMPLE MACHINES

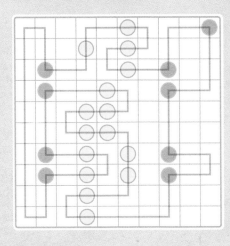

Page 168
THE SPACE ELEVATOR

	3		4			
2	3	5	1	6	4	7
6	7	1	2	5	3	4
4	6	3	7	1	5	2
7	5	6	4	3	2	1
3	4	2	6	7	1	5
1	2	7	5	4	6	3
5	1	4	3	2	7	6

5 ... 4 (left side)
3 3 6 (right side)
6 ... 4 3 1 2 (bottom)

Page 167
DYSON SPHERES

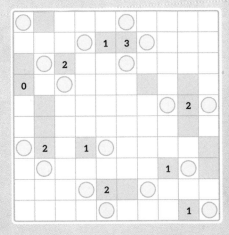

Page 169
THE SECRET OF CEMENT

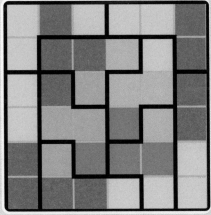

SOLUTIONS

219

Page 171
DIODES AND TRANSISTORS

Page 173
THE TURING TEST

Page 175
HOW MUCH DOES AN E-MAIL WEIGH?

WEIGHTS

 3 kg

 2 kg

 7 kg

 4 kg

Page 176
BIOMIMETICS

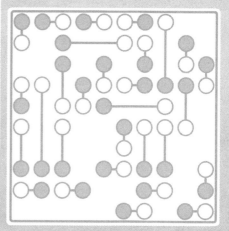

Page 177
NUCLEAR FUSION

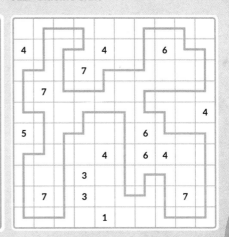

Page 181
SELF-DRIVING CARS

Page 179
HYPERLOOP

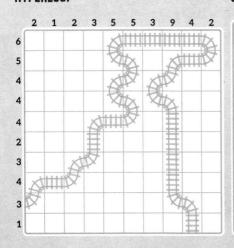

Page 183
CAN VENICE BE SAVED?

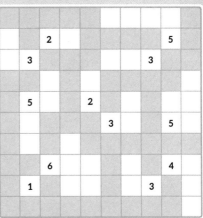

THE TECHNOLOGY QUIZ

1. What kind of engine was pioneered by Frank Whittle in the 1930s? **Answer:** jet engine

2. What name is given to the positive electrode in an electrolytic cell? **Answer:** an anode

3. What does FM stand for in radio? **Answer:** frequency modulation

4. What is the meaning of the acronyms AC and DC? **Answer:** alternating and direct current

5. What lifesaving piece of safety apparatus was invented by Humphry Davy in 1815? **Answer:** a safety lamp for miners, aka the Davy lamp

6. Digging started in 1881 on which great civil-engineering project that would not be formally opened until 1994? **Answer:** the Channel Tunnel

7. The suffix "rtf" in a computer file name stands for what? **Answer:** rich text format

8. Where might you find woofers and tweeters? **Answer:** in a speaker

9. By what name is the vehicle initially known as a landship now better known? **Answer:** a tank

10. What type of pump is named after Archimedes? **Answer:** the Archimedes' screw

11. What is unusual about a Wankel engine? **Answer:** it is rotary

12. What was the name of the first electronic digital computer, developed by the British in 1943? **Answer:** Colossus

13. Which simple machine multiplies force by rotating around a pivot, allowing the points of application and distribution of force to be separated? **Answer:** a lever

14. What kind of current does a polyphase brushless motor run on? **Answer:** AC

15. What force keeps airplanes aloft? **Answer:** lift

16. Which phenomenon did Michael Faraday elucidate with his work involving magnets, wires and batteries? **Answer:** electromagnetism

17. What is the element most commonly used in batteries for applications such as mobile phones and electric cars? **Answer:** lithium

18. A Stirling engine is powered by what sort of energy? **Answer:** heat

19. What name is given to an interconnected series of real or virtual artificial neurons (units that mimic the action of nerve cells) used for computation? **Answer:** neural network

20. Which of the following is NOT a type of logic gate used in electronic circuits? **Answer:** b)

▶ The Colossus computer which was developed by the Bletchley Park codebreakers during the Second World War to help with their cryptanalysis.

PICTURE CREDITS

The publishers would like to thank the following sources for their kind permission to reproduce the pictures in this book.

All images © Shutterstock except:

Page 10: (top left) Public Domain; 10: (bottom left) Public Domain; 10: (bottom right) Granger/Shutterstock; 12: (bottom left) Ron Frehm/AP/ Shutterstock; 56: Victor Habbick Visions/SPL; 65: Mosa'Ab Elshamy/ AP/Shutterstock; 87: Public Domain; 109: NASA/NRAO/AUI/NSF/ NASA/STScI/JPL-Caltech; 116: (bottom right) Universal History Archive/UIG/Shutterstock; 118-119: Getty/The Asahi Shimbun / Contributor; 124: (top left) Alexander Roslin: Svenska, 1775, Nationalmuseum, public domain; 144: NASA; 156: Aidan Sullivan/ Mail On Sunday/Shutterstock; 166: (top left) AP/Shutterstock; 171 (bottom) Underwood Archives/UIG/Sgutterstock; 172: (top left) Fine Art Images/Heritage Images/Getty Images.

Every effort has been made to acknowledge correctly and contact the source and/or copyright holder of each picture. Any unintentional errors or omissions will be corrected in future editions of this book.